Gender Relations in
Forest Societies in Asia

Gender Relations in Forest Societies in Asia

Patriarchy at Odds

Edited by

Govind Kelkar
Dev Nathan
Pierre Walter

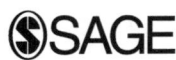

Los Angeles | London | New Delhi
Singapore | Washington DC | Melbourne

First published in 2003 by

 SAGE Publications India Pvt Ltd
B1/I-1 Mohan Cooperative Industrial Area
Mathura Road, New Delhi 110 044, India
www.sagepub.in

SAGE Publications Inc
2455 Teller Road
Thousand Oaks, California 91320, USA

SAGE Publications Ltd
1 Oliver's Yard, 55 City Road
London EC1Y 1SP, United Kingdom

SAGE Publications Asia-Pacific Pte Ltd
3 Church Street
#10-04 Samsung Hub
Singapore 049483

Published by Vivek Mehra for SAGE Publications India Pvt Ltd, typeset in 9.5/11.5 Leawood at C&M Digitals (P) Ltd., Chennai and printed at Avantika Printers, New Delhi.

Second Printing 2016

Library of Congress Cataloging-in-Publication Data

Gender relations in forest societies in Asia : patriarchy at odds / edited by Govind Kelkar, Dev Nathan, Pierre Walter.
 p. cm.
 "In collaboration with International Fund for Agricultural Development, Asian Institute of Technology [and] Center for International Forestry Reserve."
 Includes bibliographical references and Index.
 1. Rain forest people—South Asia. 2. Sex role—South Asia. 3. Women in forestry—South Asia. 4. Women in sustainable development—South Asia. 5. Matrilineal kinship—South Asia. 6. Partiarchy—South Asia. 7. South Asia—Social conditions. I. Kelkar, Govind, 1939-11. Nathan, Dev. Ill waiter, Pierre (Pierre Gilbert)

GN635.S57G45   305.3'0954—dc21   2003   2003014697

ISBN: 978-07-619-9783-2 (HB)

The SAGE Team: Sunaina Dalaya, Sushanta Gayen and Santosh Rawat

CONTENTS

I
HISTORY AND MYTH

II
FOREST MANAGEMENT

LIST OF FIGURES AND TABLES

Figures

Tables

FOREWORD

IFAD's field experience and analyses have shown that poverty in Asia is concentrated along two dimensions: geographic and social. Geographically, it is concentrated in less favored areas, such as remote uplands and mountains, marginal coastal areas, and unreliably watered drylands. Socially, it is concentrated among women, indigenous peoples, and the socially excluded, such as the *dalits* (Scheduled Castes) of India.

IFAD's interest is particularly keen where these two dimensions intersect—for example, the plight of indigenous peoples, who are the forest dwellers across Asia; and the plight of women, whose role as agents of change is a key element of IFAD's poverty reduction strategy.

In order to understand the changes taking place in the economy and social system of forest dwellers, IFAD provided funding to the Asian Institute of Technology (AIT) in Bangkok to study developments in gender relations in forest societies in Asia. The study was part of a research project entitled "Creating Space for Local Forest Management" conducted by the Center for International Forestry Research (CIFOR).

The driving force behind IFAD's assistance to the project, and the resulting articles that comprise this book, is twofold: finding ways of linking improved livelihoods with improved community management of forests; and finding ways of enhancing the social, economic and political inclusion of women in the communities and societies in which they live.

The book covers a broad range of societies in Asia (China, India, Thailand, and Malaysia). It also addresses relevant issues of history and myth, as well as contemporary developments. In so doing, it seeks to provide a broader understanding of contemporary changes in the economy and society of indigenous peoples. This in turn can lead to informed and improved policies and enhanced project interventions.

As the chapters in this book point out, there is a marked deterioration in the position of women in these indigenous communities. The deterioration is linked to the privatization that accompanies the market and to state policies, both of which have been largely based on the predominant, if not exclusive, role of men in most sectors of society. Along with the observed negative changes and impacts, the chapters also point to growing women's resistance to this status quo. Women are pushing for more "space," particularly through new forms of community and private management.

It is precisely this enhancement of indigenous women's agency that IFAD supports in its quest to contribute to the Millennium Development Goals of achieving gender equality and halving rural poverty by 2015. Undoubtedly, the two goals are inextricably linked.

Phrang Roy
Assistant President
IFAD, Rome
April 2003

Acknowledgments

The research which resulted in the articles collected in this book was funded by the United Nations, International Fund for Agricultural Development (IFAD), Rome. The research on "Changes in Gender Relations in Forest Communities in Asia" was carried out by the Gender and Development Studies of the Asian Institute of Technology (AIT), Bangkok. The study was part of a research project titled "Creating Space for Local Forest Management" conducted by the Center for International Forestry Research (CIFOR), Bogor. In China our collaborating institution was the Institute of Ethnology, Yunnan Academy of Social Sciences, Kunming. The chapters in this book were all originally published in the journal *Gender, Technology and Development*, Vol. 4. No. 3, and Vol. 5 Nos. 1, 2 and 3 (Sage Publications, New Delhi). We are grateful to AIT, for permission to reprint these articles.

We extend our appreciation to all these institutions for supporting us and helping us carry out the research in this book. Obviously the institutions involved in this research project are not responsible for the analyses and opinions expressed in the articles, which remain that of the authors' alone.

The editors would like to thank the following individuals for their support and comments in the course of coordinating, carrying out, and preparing for this book: Phrang Roy, Assistant President, IFAD; Ganesh Thapa, Regional Economist, IFAD/Asia and the Pacific Division; Lini Wollenberg and David Edmunds of CIFOR; A. Contreras, Liu Dachang, Madhu Sarin and Neera Singh, the country coordinators of the CIFOR research project; Gajendra Singh, one of the principal investigators of the AIT project; and Guo Dalie, Director of the Institute of Ethnology, Kunming. This book would not have been possible without their support.

The editors would also like to thank the staff of Gender and Development Studies, AIT, who cheerfully and efficiently administered a research project with so many researchers located in various countries of Asia. In particular we would like to thank Yu Xiaogang, Veena N., Emilyn Madayag, Agnes Pardillia, and Girija Shreshta.

INTRODUCTION

Forest Societies in Asia: Gender Relations and Change

GOVIND KELKAR AND DEV NATHAN

Introduction

Of the thousands of studies of forest management in Asia, only a tiny proportion mentions the role of women or pays attention to gender relations. Projects are often designed in terms of the household or communities in which men are the household heads and community decision-makers and owners or managers of forestlands. Some writers, however, see gender relations as a crucial factor in the management of land and forests and point to the continuing invisibility of women (Bosu Mullick, 2000; Kelkar and Nathan, 1991; Sarin, 1996, 2000; Singh, 1999; Townsend, 1995; Wickramasinghe, 1999). This invisibility compounds poverty, shortages of food, fodder and fuel, and the greatly increased workloads of forest-based women.

An initial intellectual curiosity drew us to study forest-based societies in which women were/are not transposed into extreme inequalities in social, political, and economic spheres. We also looked into the manner in which gender relations were transformed in the shift from matrilineal to patrilineal societies, within forest-based communities. If male domination was not always the rule, how did gender/social relations change from societies without male domination to ones where male domination was presented as the norm?

Most studies of matrilineal societies have focused on one of the two issues: the division of authority between women and men (Bachofen, 1861; Richards, 1950; Schlegel, 1972), and the conditions that give rise to and sustain matrilineal systems (Divale, 1984; Gough, 1959; Mann, 1987). Undoubtedly, these studies are significant to our understanding of the development of a patriarchal gender system, and of the consequent feminist demand of gender equality. We would like to make only one theoretical point here: many studies of the gender system in indigenous societies in Asia dispute the view that men enjoy superiority over women universally. Also, these studies further suggest: (a) that the matrilineal systems were characterized by harmony (free of domestic and community violence), and reproductive power associated with women; and (b) women were the source of wisdom and technology, that women brought about all the animals, plants tools and techniques that constitute this world.

We launched 12 studies of forest-dwelling indigenous peoples in Asia. Of these, six studies were done in matrilineal communities, five in patrilineal communities, and one on women and hunting.[1] These case studies were conducted to address the following major questions: How do gender relations within and outside the household affect use and management of forests? What is the extent of women's centrality in provision of livelihood, through an analysis of their role and status in management of natural resources, with special attention to forests? What can we learn about gender relations from forest-dwelling societies that are characterized by cultural valuation of women's economic, political and ritual roles, and absence of institutionalized male control? Does the structure of gender relations within the household and in the community change as members respond to broad religio-cultural, social and economic restructuring of the indigenous societies?

While we began with a broad set of questions, throughout our research we refined our questions based on the emerging field data. We sought to describe and explain gender relations in forest societies through the voices of indigenous women and men. The 12 studies used open-ended individual and collective interviews, and fieldwork-based qualitative assessments in their overall analysis. Moreover, we had several meetings and mid-research workshops to discuss fieldwork problems and the nature and quality of our inquiries. We visited several field sites, e.g., villages in Meghalaya, India, Tamang villages in Nepal, Rungus villages and longhouses in Sabah, Malaysia, and Mosuo and Naxi forest communities in Yunnan, China. We found our research teams' work engaging and important. They took us to field sites, arranged individual and collective interactions with local women and men, and created a unique opportunity to transform us from "foreign researchers" to multisited critics with the task to promote indigenous women's advancement, and assist policy-making in halting the forest world's too invisible inequalities. Our analysis, therefore, is not based on a presumption of meta narrative but on a methodological requirement to consult with the silent voices of indigenous women and men. Further, we attempt to do something more than deconstructing scholarship on matrilineal societies simply by retelling indigenous women's stories about resistance to male dominance.

We looked at changes in gender relations in forest societies in four situations. The first is that of the imposition of colonial or national state rule over forest communities, and the takeover of forests for central purposes. Along with the exclusion of local communities from the forest, this also resulted in the imposition or at least importation of "mainstream" values that restricted women's space and exalted the domestication of women.

The second situation is that of the revolts, historical and contemporary, to reestablish local community control over the forests. What is notable is that a number of these revolts were accompanied or preceded by "internal purification" movements to eliminate women who were denounced as witches.

The third situation is that of the response of national states to these autonomy movements by shifting to devolution as a policy. Of course, this is accompained by the fiscal problems of these central states, and the failure of state enclosure, which made such devolution necessary and feasible. The impact of these contemporary devolution measures was, in the first instance, usually negative for women's participation. Women's exclusion from traditional village councils was carried over into formal state-sponsored committees.

The fourth and current situation is where women's inclusion in committees is becoming more a policy norm. In many places, all-women groups have come up for forest management and protection. Women are seen to perform better in many management and production tasks. However, these new norms of women's inclusion, though still limited in space both vertically and horizontally, have also come about through a process of struggle by women, often supported by various external actors.

Gender Relations in Forests Societies

Gender relations are complex, dynamic and socially embedded having many interlocked dimensions. Cultural traditions of women's exclusion from community management confers authority and prestige on men. Men hold virtually all formal positions of power and decision-making in villages under patrilineal systems, though women often exercise considerable influence in certain areas of village life. In matrilineal systems, women have an especially effective, indirect power in maintaining the lineage and therefore owning children. They have rights over ancestral property, and control and knowledge of ritualistic activity, including being the spiritual heads of the community, for example, *Syiem Sad* among Khasi in India and *bobolizan* among Rungus in Sabah, Malaysia.

In most patrilineal societies, women's major responsibility in reproduction and/or income-earning does not necessarily lead to social empowerment or gender equality within the household (Chhakchhuak, 2000; Kelkar and Nathan, 1991; Munshi, 1986; Yang, 2000). However, the presence of rights to the forest, and women's rights to access forest resources can mitigate, to some extent, this inequality in gender relations. K.S. Singh (1999) and Bosu Mullick (2000) found that in forest-based (patrilineal) communities, because of their involvement in gathering from forests and their marginal dependence on agricultural produce, women are economically more independent and have a higher status than their counterparts in the rest of India. Among the patrilineal Naga of northeast India, for example, women's role in swidden agriculture and in the processing of forest products for sale has kept gender relations relatively balanced.

In the more male-dominated society of Enzong village (a mixed Naxi and Han village, 8 kms from the town of Lijiang) women have no right to forests, land or trees, all of which are inherited from father to son, and forest distribution is carried out on the basis of the (male) head of household. A woman, after marriage, acquires access to her husband's forest. Furthermore, women are allowed neither to climb trees nor to cut trunks even if these are needed for house construction.

Spirits were supposed to reside in trees and a menstruating woman might pollute the tree and thus bring down the wrath of the forest spirit.

The intimate connection between the establishing of new family units and the fireplace exists widely among the indigenous people in southwestern China (Yang, 2000). In many cultures, the fireplaces are classified into female and male fireplaces, each with its own name signaling gender divisions and gender hierarchies. Among the Wa people, married women and men live in separate household spaces in a family compound. Gender mingled living is not allowed even for the aged couples as a rule. The fireplace in the women's room is for cooking and boiling water while that of the men's room is regarded as the residence of the God of the fireplace, which is worshipped very often. The men's fireplace also serves as a guest reception area in addition to being used to supply daily heat and boiling water for the men. By contrast, among the matrilineal Mosuo, women have their own residence to entertain men and the latter have no permanent place to stay but their mother's homes. The senior women in the Mosuo household manages family life, and therefore is seen to deserve a superior seating at the fireplace.

What we are dealing with is a whole complex of social behavior, of practices that constitute incorporated social or gender relations. The complex of practices deals with knowledge, with religious rites and symbols, marriage and inheritance systems, economic rights, control over sexuality and reproductive powers, norms of behavior, accepted forms of social excellence and ability, and differential access to social knowledge—all of which together constitute gender relations.

Still, in matrilineal systems, women's control over forestlands has generally enhanced gender equality by giving women a greater say in how forestlands are used. In the Chota Nagpur villages of central India, Samar Bosu Mullick (2000) notices present day practices that socially acknowledge women's knowledge of forests and agriculture. When the Munda (the headmen) go from one village to another, their wives lead them. Women's knowledge of seeds, herbs, and plants is considered precious both in the family and community. Their knowledge of the roots of a particular plant is used to brew rice beer, the most sacred and popular drink of the people. Their role in the preparation of cultivable land is also very important; they are seen working with men in field preparation and reclamation of forestland. Women's contribution to the development of agriculture is further confirmed by the 'myth of the preparation of the first plough.' The

Supreme Being's wife is described as the real inventor of
the technology of plough-making. Thus women's right to the newly-
reclaimed forestland and its produce received a permanent place in
the customary law of the Munda people.

Matrilineal societies often associate women with forests through
specialized roles in healing and religious ceremonies. Indra Munshi's
interviews with the Warlis in India suggest that many women have a
fair knowledge of medicinal properties of trees, roots, herbs, and
medicinal plants useful for reproductive health, childbirth, abortion
and so on. Common ailments are treated at home. In the Tamang village Chisapani in Nepal, Suman Subba finds "The women shamans
are accepted as equally knowledgeable and powerful as the male
shamans" (2000: 9). There is no gender difference in carrying out rituals and healing people; except that the women shamans do not sacrifice animals and ask for help from a male member of the family. The
women having a "soft heart" was the stated reason for this difference
in the sacrificial ceremonies. Both women and men, however,
acquire the knowledge of shamanic treatment of diseases. This can
be acquired either by learning from the senior shamans or through a
spiritual knowledge relationship with a *Ban Jhankri* (the forest
shaman), a female or male spirit who stays in the forest—in its trees,
springs, caves, cliffs, hills etc. Also, this knowledge can be acquired
through spiritual transference into a new body, usually done at the
deathbed of a senior shaman (grandmother or grandfather), to a
member within the family. In either case, women's role in spiritual
life, associated with knowledge of forests, places her in a position of
relative power within the household and the community.

Male domination in the Han political system introduced the new
perception that it is men who control both social knowledge and
family resources. Both the Han and the Tibetan Lamaism became the
feature of the male dominant gender system. However, in traditional
Mosuo society, Lamaism was the single most powerful challenger to
the Mosuo matrilineal ideology and gender constructs. While the
Han Chinese bureaucracy "tried to 'civilize' the Mosuo largely
through administrative measures, and without much success so far,"
Lamaism took on a much more subtle approach; "first it diffused the
idea of the female sun and then gently promoted the idea that the
father is like the sun, and most likely also reinterpreted the meaning
of the female sun of the Mosuo" (Shih, 1993: 178). Lamaism thus consequently "transformed the Mosuo conception about maleness and
femaleness" (ibid.). During 1993 and 1997, when we did our fieldwork

in the Mosuo region, we saw that any household having more than two sons would have at least one who would be Lama. The Lama, mostly lived in the upper part of the house, did not do any kind of physical labor, not even cooking for himself. His food was prepared by his sister or mother or brother, and sent to him in the upper chamber. Likewise, with the advent of modernity and Hindu influence the number of women shamans possessing traditional knowledge among the Tamang of Nepal has been sharply reduced over the past few decades. There were three major reasons for the increased social differentiation between women and men: women who possessed such knowledge were treated as potentially engaged in witchcraft; women's dignity was reduced within the family; and, the burden of household duties was increased. In some cases the family would restrain the woman from working as a shaman, due to fear of social ostracism.

Thus, before the advent of state pressure on matrilineal societies, gender relations were relatively equal. Based on women's role in production, their special knowledge of forests, and their place in the cultural and religious life of matrilineal communities, women enjoyed considerable space within the household and the community to make decisions about resource use. Unfortunately, maintaining this position of power has been difficult, particularly in the face of pressures from the state in favor of centralizing greater partriarchy.

Expanding State Space into Forest

State-sponsored colonization by the dominant religio-cultural regimes like Hinduism, Christianity, Lama Buddhism and Confucianism are noted as having had a destructive effect on indigenous gender relations. They disordered the position of women as that of subordinate to men, and reinforced their exclusion from political and spiritual life and community decision-making. Among the Naxi, for example, the matrilineal system was abolished and replaced by patrilineal inheritance; marriage for love was discouraged and replaced by arranged marriage; and the Confucian values of a woman being subordinate to her father, husband and son were promoted. The effect of these measures was to deny women opportunities to participate as full members of the forest community. Not surprisingly, there is an element of local male support in this colonization, for the men seemed to have acquired power and representation in the dominant society's image of indigenous people. Indigenous women's argument that their traditional roles included full participation of women in many aspects of spiritual and political decision-making has been ignored not only by the alien authorities, but by their fathers, husbands, and brothers as well.

For many forest dwellers, the major agency of change in their cultural system and gender relations was colonial education, which came through missionary or religious schooling, followed later by public and secular schooling. These measures were designed to deny women the opportunity to participate as full members of the community.

During the expansion of Christianity among the Rungus in Sabah in 1952–59, for example, the missionaries would directly deal with the village headman which left the *bobolizan* (a woman priest and healer, having absolute power in making decisions which have to be carried out by the community) in the dark on what the consequences were for their traditional religion. Further, the missionaries set up three types of schools: primary schools, farm schools, and domestic science schools, and also dispensaries for free medical and health care. Domestic science schools channeled women into domesticity, denying their past role in productive and political life, and limiting its future. Free health care undermined the role of the *bobolizan* as healer, and consequently her status and political power. "The methods used by the missionaries directly reduced the influence of the *bobolizan* in the community" (Porodong, 2001: 88).

Likewise, in the Khasi Hills of India, Nongbri pointed out that Christianity and education have not only altered the value of forest, land and land relations but also "seriously eroded the ideological and material basis" of Khasi matriliny, it has "helped create a social milieu fostering patriarchal values. As an agent of modernization, Christianity propagated western patriarchal values even to the non-Christian population through its educational system" (Nongbri, 2000a: 383–84).

The effect of such modern schooling was to silence and suppress the traditionally acknowledged women decision-makers, and the indigenous ways of knowing, their language and rituals for knowing. In the name of western/modern knowledge, development and a "civilized culture," emerged male power and the hierarchy of gender relations. In her study of Shu, the Naxi Nature Goddess, Xi Yuhua shows that embedded matriarchy was a character of the traditional

Naxi society. Shu were predominantly feminine with masculine characteristics, having a three-in-one image; frog head, human body, and serpent tail. As time passed, patriarchy surpassed matriarchy and the conscious/unconscious forms of Shu began to be replaced by that of masculine elemental character.

At the same time, colonization of indigenous societies generally led to centralization of forest management as well. Even before the advent of European domination, indigenous peoples saw their forests come under the management of outsiders in India, China and much of the rest of Asia where princes, kings, and emperors reigned. In China, the area of Lijiang was brought under direct Han Confucian rule in 1792. This policy, among other things, brought the forests belonging to the former chief of the Naxi under the central state control. The Han also introduced techniques of tree management, like grafting, and promoted conversion of forests into orchards (Guo Da Lie, 2000). The arrival of European foresters simply introduced further restrictions on indigenous peoples' use of forests.

By centralizing forest management, states weakened an important source of women's power in matrilineal societies. When forests were under local control, particularly in matrilineal societies, women played an important role in forest-based production, and often enjoyed a high status based on their knowledge of forest flora and fauna, and their role in religious rituals with strong ties to the forest. While women certainly continued to use forests after centralization, they often had to do so clandestinely and in short visits. In addition, many forests were changed into monocrops that provided few of the resources that women controlled historically. With limited access to a much-altered forest, women's ability to fend off forces of patriarchy was greatly reduced.

Gender Struggle in Forest Autonomy Movements

State efforts to centralize forest management did not go unopposed. Widespread rebellions against the loss of local control over forests were seen, particularly in India and China (Bosu Mullick, 2000; Kelkar and Nathan, 1991; K.S. Singh, 1999). Yet these movements did not often reassert women's equal rights with respect to forest management, or any other aspect of social life for that matter. A shift in gender power from women to men was already well underway when such movements started, and local men used the moment to further consolidate patriarchy. In the process of changing forest use, from swidden systems to settled, privately-owned and inherited rice fields, and the change from community access to private access to forest products, women had lost the source of their power and status. Men were fighting for the return of forests, not gender equality.

Historical and contemporary movements for local control over forests and political autonomy among communities in Jharkhand, India—the Santhal, Munda and Ho—are a case in point. Before the beginning of every rebellion or movement, the men undertook a special drive to "cleanse" society by eliminating witches and poison givers, "the dirtiest creatures who keep evil spirits." The mid-19th century Santhal rebellion was preceded by and accompanied the killing of women denounced as witches. In the Birsa Munda-led movement to secure recognition of the clan brotherhood's right to the forests they had settled in, women were denounced as witches and killed for superstitious practices like witchcraft (K.S. Singh, 1999). As Bosu Mullick (2000) points out that the areas where the contemporary and ongoing Ho movement for control over forests and political autonomy are strongest (Khuntpani and Tonto Blocks of Singbhum district) are also the areas where the persecution of women as witches is most prevalent.

The pervasiveness of the demonization of women as witches (givers of poison or *najom* among the Ho, *dain* or witch among the Santhal and Munda, *yanggu* or nurturers of poisonous insects among the Naxi, *pippa* among the Dai, Shan and Tai-speaking peoples in the forests of peninsular Southeast Asia) or of the overcoming of the female by the male power (the stealing of the secret of salt among the Lua in north Thailand, or the killing of the most accomplished *bobolizan*, woman priestess, in order to attain the highest rank of warrior among the Rungus of Sabah) show that this gender struggle is both internal to these communities and related to some important historical change. This strengthening or imposition of male power was and is seen as a necessary step to enable forest-based communities to regain control of their forests and overall livelihoods. (For detailed accounts see Bosu Mullick, 2000; Cholthira, 2001; Kelkar and Nathan, 1991; Munshi, 2000; Nathan, Kelkar and Yu, 1999; Porodong, 2001; and Xi Yuhua, 2000.)

The categorization of women and men into witches and witch-hunters, respectively, was an essential part of the process of establishing the authority of men. The denunciation of ritual knowledge by women as evil can be seen as an attempt by the denouncers (men) to change the established order (which was very likely one of joint authority of women and men), and to set up in its place an order based on the sole authority of men. Thus, what is seen as a resurgence of the forest community is also simultaneously a struggle to establish patriarchy.

Women's Voices of Resistance and Demand for a Role in Forest Management

To describe indigenous women as silent observers of the male appropriation of traditional power and resources oversimplifies both the women's voices of resistance and the range of ways in which they have expressed resistance. There are indigenous women who would speak publicly about the growing male dominance and control of resources, even as they challenged central control over the forests they depended on. In fact, women have faced struggle on two fronts, against patriarchy within and outside their own communities, and against the seizure of forests that were once their source of power and authority, and on which they still often depend for their livelihoods today.

Of course, many women remain silent, but they would speak if power and resource inequalities did not create obstacles. Some choose not to speak publicly but instead exercise informal resistance in what James Scott describes "off stage defiance" (1985: 23), "infrapolitics of subordinate groups.... A wide variety of low profile forms of resistance that dare not speak in their own name" (1990: 19). There are also women who keep quiet and show nothing but compliance to male dominance; this is done as a result of fear of insult and physical assault. Moreover, gender-based domination is complex. In the case of women, relations of domination have typically been both personal and community-based; joint reproduction in the family and home without any control over resources has meant that "imagining an entirely separate existence for the women as a subordinated group requires a more radical step than it has for poor peasants, working class or slaves" (ibid.). It is not surprising, then, that women do not publicly speak out against their oppression and subordination, since like any other subordinated group, they may be socialized into accepting a view of their position and interest as prescribed from above, in maintaining the male hierarchy in gender relations.

Yet, many of the women we spoke with recognized the need to struggle on two fronts. First, against the social authority of men, and women's exclusion from participation in the organization of their community. Second, against the encroachment of the state on forests and other resources of importance to their communities. The indigenous women expressed anxiety about physical and cultural survival of their communities, while also being critical and challenging growing male dominance.

In an interview with two women school teachers in an east Khasi Hills village in March 2000, we learned that in many cases in recent years, *khadduh* (the youngest daughter who has the obligation to provide support and succor to all members of the family) has begun to assert her claims to full management and ownership rights of the parental property. Conditions for such a claim came up when the uncle (the mother's brother) or the brother of *khadduh* stealthily but unsuccessfully signed away the family land and/or trees for his personal benefit.

Indigenous women across matrilineal and patriarchal or patrilineal systems of Asia have stories to tell which not only question the assumed nature of the male authority and practices it generates, but also serve to tell an alternative story, the history of all forms of injustice and the history of their resistance through the eyes of women. A newspaper, *The Shillong Times* of northeast India, carries a popular column by Patricia Mukhim every Friday. Questioning the Khasi "wealthy, powerful men of modern generation" who find the Khasi matrilineal system "somewhat unacceptable," she says:

These women have never seen a Khasi *Dorbar* in progress. It is in the *Dorbar* that one can assess the strength of women. What kind of power can women exert if they are not allowed to speak and make a point in the *Dorbars* ... the male chauvinist [is] trying to protect his position. The *Dorbar* is today the only domain of male superiority. The Khasi male has always considered women to be a little less intelligent, if not outrightly scatterbrained (February 19, 1999).

Another Khasi woman, a social analyst and historian, Tiplut Nongbri criticizes the decline in the authority of the Khasi spiritual head, the

Syiem Sad, "the source of women's power and influence in the society." In present times, both in religious and political matters the powers of the *Syiem Sad* are subject to a number of checks and constraints (2000a: 374–75). In a similar vein, a Naxi woman sociologist, He Zhonghua protests against the decline in the position of Mosuo women in Yunnan, China: that women are socially defined to be of poor ability with little education and incapable of participating in community affairs. In reality, however, she notes, "some of the women are not less qualified, and some of the men are not more educated, yet the seats of the village heads or officials are always for those said to be abler. No opportunities for women!" (ibid.).

Khasi, Mosuo, Rungus, Warli and Santhal women do not, therefore, necessarily share their men's worldview about women or social and technological change. The women are critical and resist not only their subordinated position, but also the conversion of natural forests into commercially useful monocultures and the overall depletion and degradation of forests. Indra Munshi (2000: 7) describes Warli women's voices against the government control over forest resources, "When *adivasi* (indigenous) women go to the forest, they are threatened and chased by the forester.... Women are tired of roaming in the forest, and even then they don't get fuel wood.... The government has finished the forests, how can we get wood?"

If women are excluded from forest management, the devolution through community-based forest management does not make any real difference to women's lives. The real question for them is: how inclusive is the management of forest resources? While narrating their experiences related to forest use at a workshop in Orissa in 1993, village women's response was, "Sir, what can we speak about forest, what is there to say about forests, forests used to be guarded by the forest guards earlier, now it is guarded by youth club groups. For us it makes no difference." There has been "only a shift in the *danda* (staff)" from the hands of the forest guard to those of the local youth (Singh, 2000: 20).

The Chipko Movement in the northern hills of India is justly famous for its picture of women resisting logging. In many self-initiated community forest management schemes in Orissa, women took the lead in protecting the forest. There are good reasons for women's initiatives in this area. Since it is women's responsibility to provide for the family, and particularly to collect fuelwood and water, they are immediately affected by the loss of ecological services, like water, that are dependent on proper forest management. As women have protested the loss of forests and forest access, sometimes along with men, but also often on their own, states have responded with various measures to devolve forest management to the local level. The question is, to what effect?

Devolution Policy and Gender Relations

The local and regional subordination of women in forest societies today is unquestionable, even in what we consider as matrilineal societies. The historical development of forest policy prior to devolution is known for the organized power of forest guards (mostly male) and absence of women in all traditional and formal institutions of forest management. Forest-dwelling women, however, have the primary responsibility for collection of food, fodder, and fuel for the family from the forests. To what extent has devolution changed these gender relations? Has it changed the forest policy from being class-based and andocentric to community-based, participatory, and inclusive in the real sense of the term—devolving authority to forest-dwelling women as well as men?

Many devolution policies target the community as the unit to make decisions once made by the government forest department. This is certainly true for joint forest management (JFM) in India, community-based forest management in the Philippines, and some of the collective management policies in China. If women are excluded from community decision-making, however, devolution through community-based forest management does not make any real difference to women's lives. Where women are involved in decision-making about forest, they tend to take account of the needs of fuel and fodder, which are otherwise ignored by men sitting on forest protection or similar community-level committees. The Nepal Leasehold Forestry, for example, targeted women's user groups specifically, included women's needs of fuel and fodder, and as a result has had a substantial impact on women's lives. In contrast, the men-dominated Forest User Groups in Nepal have not (Subba, 2000).

Certainly, women have played a variety of roles in community-based forest management. Yang Fuquan's (2000) study of Naxi in Longquan village in Yunnan shows that some women were elected to important roles in protecting the forest resources of the community. Importantly, for the post of forest or mountain guard, villagers will elect a person who is honest and frank, and can act justly, and it is the choice of the village women that counts in this appointment. A forest guard has a specific privilege: the dying trees of the collective forests are considered as her property, and she can take them back to her house. If she confiscates the axes, machetes and sickles of people who fell trees or otherwise violate village regulations about the use of forests, she can take the money or gifts that people pay her to redeem their axes, etc.

Box 1

An 80-year-old woman, who had been a forest guard for 22 years, from the mid-1950s onward, summarized her experiences of looking after the forest resource:

I, as a woman, could be successful in managing the public mountain of two villages for a long time. It is because (the people of the two villages) backed me up so that I could boldly manage the collective forests of the two villages.

The villagers trusted her immensely and gave her nine work points per day, which was the highest payment at that time. During the interview, the other male forest guard and villagers added: "She is a courageous and robust Naxi woman. When she was the forest guard, many men were afraid of her and also respected her." They further acknowledged that women are the backbone of the family and their working time is 10 to 20 percent more than that of men. The women, however, said that their work was actually 20 to 50 percent more than that of men, and men subsequently agreed.

There are also examples of community forest management in India where women have played an active role in initiating forest protection, and several cases where women's committees (Mahila Samitis) are managing forests. In Baghamunda village in Orissa, the Mahila Samiti took over the forest protection and management responsibility in 1998 after the local youth club proved to be ineffective. The Mahila Samiti deploys five members on a rotational basis for guard duty everyday. The women combine their guard duties with household responsibilities of collecting fuel, fodder, and other forest products (Vasundhara, 2000). It is important to note that village committees tend to treat this women's protection responsibility as an extension of women's daily tasks of fuel and non-timber forest products (NTFP) collection from forests. However, when it comes to the inclusion of women in decision-making about the management of forests, the male resistance is summed up in a statement made by a leading member of the Forest Protection Committee of Lapanga village in Orissa, "We are not so modern that we would involve women in Forest Protection Committee" (ibid.).

In the Xiang forest station in Lugu Lake, China, He Zhonghua learned "In afforestation, 80 percent are women, in putting out mountain fires, 40 percent are women, in planting trees, 80 percent are women. And, when fighting fires, it is easier to mobilize the women than the men." Furthermore, while cutting firewood, women choose to cut the twigs but men would cut the full grown and even young trees at random. She adds, "in fact, they [men] are destroying the forest. We saw this with our own eyes" (2001: 58). Men, however, chair all the forest rituals and religious ceremonies, and also dominate village and forest management committees. Women's interests, labor and skills are not considered in many of the decisions they make. In each case, women are prevented from gaining access to resources such as fuelwood or broom grass, which are the basis of their livelihoods, and are also excluded from the decision-making bodies that determine access in the first place. In each case, devolution did not make any substantial difference to the lives of the poor indigenous women. In many of the sites we visited, devolution targeting communities has yet to address the question of culturally-embedded gender relations and the gender-specific exclusion of indigenous women from local forest management institutions such as protection committees.

Devolution may not necessarily take the form of community management of an entire forest. Many policies seek to devolve control over specific forest products, especially NTFPs. How has devolution of this sort affected gender relations?

NTFP: The Main Collectors are Women

Devolution has failed to address the question of gender and social inequalities and, therefore, in many forest societies it has resulted in both promotion of accumulation of forest income and power under the control of local elites, and the exclusion of women from ownership, control and institutional management of forests and other resources. Hence, forest societies have experienced widening socio-economic disparities and deepening gender inequalities. Three major constraints in the development of indigenous peoples relate to (a) interventions from outside, which have by and large been extractive; (b) the indigenous peoples' own fragile production structures, further threatened by these extractive external relations; and, (c) weakening institutional mechanisms. Women are worst affected as they have little or no say in community affairs and their contact with the outside world is minimal. And when they do come into contact with external factors, these are usually exploitative.

Two facts stand out about NTFP. The first is that their sale is an important part of forest dwellers' income, and often constitute almost their entire cash income. The second is that while women are not the exclusive collectors of NTFP they are the main collectors, with a few exceptions such as cane and bamboo. In many indigenous communities women are also involved in the sale of NTFP, and where that is so, it provides them with an income that, unlike the produce of land, is not mediated by their relationship with men. These two facts or stylized facts are valid for virtually all forest-dwelling indigenous people of Asia.

Devolution in the case of NTFP has taken two forms. The first is the delineation of access rights in collecting NTFP. In this access system, those who are members of the community have access. As discussed elsewhere, this usually also excludes effective restrictions on quantities harvested. The decline of NTFP in the wild so to say, has accelerated the trend of domestication of valuable NTFP. This process of domestication, dependent as it is on ownership of land (home gardens), works against the access of the poorest women, widows, and otherwise single women. In Bastar, it was observed that such single women were the persons who depended most on collection of NTFP—sal seeds especially—from the open forest, while women from families that had substantial home gardens did not take up the sal seed collection.

The second form of devolution in NTFP has been the formation of local women's organizations to intervene in the trading process. The experience of the organizations in Bastar shows that this devolution has resulted in some livelihood benefit. The limitation of such devolution, however, lies in the bureaucratization of the process and the concomitant limited development of women's capabilities. It should, however, also be pointed out that, where some women have seized the opportunity (e.g., Asana village in Bastar, India), devolution has contributed to developing the capability of women in the market and to strengthening their bargaining position in the market for NTFP. In addition to bureaucratization, the failure to address the food security problems of the poorest producers, and the concomitant interlinking of markets for outputs (NTFP) and daily necessities, also limit the benefits that the poorest (usually single) women get from the devolution.

Privatization of Forests

Devolution is not only the result of state policy. Markets drive devolution too. Opportunities and constraints posed by market development for forest products can encourage the breakdown of community-level institutions for managing resources in favor of privatized, individual management. This form of devolution also has discernable impacts on gender relations that bear examination.

In northeast India, privatization has brought a number of changes to women's lives, with important differences by location. Among the Khasi people, the status of women has depended on their ownership of, and claims upon, ancestral property. However, the registration of former community-owned forests as private lands does not constitute ancestral property. It comes under what is now called "self-acquired" property, the right to inherit which the men were able to legislate by different principles. Where, however, forests were privatized by a village that was deciding to divide its common forests, the new property was also in the names of the women whose houses the lands were attached to.

More important than the formal changes, however, have been the changes in management. Even in the traditional system, the ancestral property of women was managed by the maternal uncle or the brother. The direct role played by the maternal uncle or brother continued even after the men married into other clans, since marriages often took place within the same village. But increasingly, it is the husbands who are effectively managing land, including forestland. More than land itself, it is capital that has become the key economic resource, and this capital as self-acquired property is passed on from father to son. The rise of the timber industry has enabled men as husbands to take control of the family's economy. Women's ownership of land is no longer the determining feature of the Khasi property system, but has been reduced to a vestigial right which, however, does place women in a better position than if they had been completely propertyless.

Among the landless Khasi families, women do not own even the house plot. The main source of cash income is wages from logging, which is earned and controlled by men. In such landless Khasi families there is clearly very strong male domination. When discussing the problems of men demanding money for liquor, women said that if they did not hand over the money they would be beaten, and even thrown out of the house. Being thrown out of the house is something that a house-owning Khasi woman is not likely to be subjected to.

The Nagas of northeast India are patrilineal communities, with women having no inheritance rights over property. The men, however, have played a very minimal role in production. Men's social role was centered around war and fighting. Headhunting and bloody village feuds have disappeared, but in most Naga areas there is a continued insurgency against Indian rule. Men who remain in the villages, however, now have a lot of spare time. Barring land preparation, men only play secondary and occasional roles in agriculture.

Besides agriculture, the other main economic activity of the Nagas is the weaving of shawls. Such weaving by women is the main source of cash. In Ukhrul district of Manipur state of India, there are numerous restrictions on the harvesting and sale of NTFP. While there is no restriction on NTFP extraction for self-consumption, sale outside the village is only permitted to monopoly traders, who secure their monopoly rights in periodic auctions. The low village-gate prices of NTFP resulting from the restrictions on sale here meant that collection and sale of NTFP is not a very lucrative occupation. Consequently, the main source of cash income is from the shawls woven by the Naga women.

The more recent introduction of formalized village management of the economy, the setting aside of earlier fallow as village reserved forests, and the associated flow of funds into the village through projects like the IFAD-funded Northeast India Natural Resource Management Project, have also served to increase men's control over the domestic economy. Though their knowledge of the local economy is very limited and they neither play much of a role in production nor in marketing of agricultural produce, their role as community managers still enables them to exclude women from community-level decision-making about natural resource management, including the management of forests.

Logging and Gender Roles

Logging is a masculine activity. The timber trade, which began mainly in the 1970s with the construction of roads for the Indian army, placed substantial amounts of cash in the hands of men. Where the growth of the timber trade encouraged privatization and a reorientation of forest production, it strengthened men's control over the domestic economy, which earlier had been almost the sole domain of women. Where devolution encouraged logging by local collectives, it reinforced male decision-making within local institutions, often to the detriment of women's interests.

In Meghalaya, this was the chief source of cash income among the landless who worked as wage laborers in logging. Unlike the case of the upper class Khasi women, the women of this class had no property in their names, not even house plots. Among this section of landless Khasi, men's domination was very strongly established through the wage earnings from logging.

For forest owners, the situation was somewhat different. Besides effectively taking control of their wives' lands (and not being just managers of their mother's or sisters' land as was traditionally the case) sawmill owners (mostly men) are also accumulating capital, based in large part on the transformation of trees on the land into capital in mill accounts. This capital then becomes "self-acquired" property, which is not subject to the matrilineal inheritance system of the Khasi. While the land, including forestland, remains in the names of women, it was effectively managed by the husbands and, more important, it was the ability to mobilize capital and not ownership of land as such, which became the key economic resource. And this capital as "self-acquired" property is passed on from father to son. Thus, the rise of the timber industry in Meghalaya has been a major factor in the demise of matrilineality.

In China, among the Naxi community of Lijiang, men, while away in logging, did not need to check with their wives before spending money on themselves. On returning too, both women and men reported that men deducted their "pocket money" (for cigarettes and alcohol) before handing over the rest of the money to their wives. Women still retained control over the rest of the money. Naxi women said that they needed to maintain this control, for men tended to spend the family money on nonessentials. The fact that men now earned the main cash income of the family made it easy for them to claim a part as their legitimate "pocket money," but this does not seem to have overturned the existing control of women over the family income. At the same time, men remained "heads" of the family, a function reinforced by the distribution of village income from logging (the village collection charges) to family heads and not to individual persons. Furthermore, among the Lisu people, where men had traditionally maintained control of the purse, the growth of logging only reinforced this role.

Women saw their role as having become more important, since they now had to perform some of the jobs that were formerly done by men. For instance, men were away during the crucial period when floods threatened the fields. The task of flood protection then had to be undertaken by women on their own. Further, logging led to an overall shortage of male labor supply in the area. A woman and a man, with the woman leading the buffalo, now did the ploughing which was formerly done by two men.

While women reported that their confidence grew on taking up new tasks, this also meant an increase in the women's already heavy burden of work. Among the Naxi, the route to economic prosperity is said to lie in acquiring a "hard-working wife." The increase in income did not lessen the women's burden. At the same time, men who often had little labor to perform, found themselves pretty busy during logging time. The traditional masculine task of cutting trees (and constructing houses) became the main source of cash income. This made the sharing of the labor burden between women and men somewhat more equal.

Logging provided an avenue for men to make a substantial contribution to the family cash income. Except for some amounts spent on liquor and cigarettes (amounting to about 20 percent of that left after meeting food and board costs), the rest of the money was handed over to the women to run the family economy. This was a big change from the traditional situation where men contributed relatively little to household cash.

There was, however, little change in the expectations from women and men, and the socialization on the basis of accepted gender roles. Women still had the main responsibility of maintaining and sustaining the families. Young men were not expected to contribute to the family particularly before marriage. Some young men who went to towns to seek work did remit money home, or maintain younger siblings in school, but most young men did not contribute to their families; rather they regularly sought some money from their families, as mentioned by Naxi mothers in Xitoh township. On the other hand, all young women who went to Lijiang or other towns to seek work regularly remitted money to their families.

These expected gender roles that play such a strong role in socialization had their influence on children's continuation in school and higher education. Naxi mothers stated that if they could not educate both children, they would decide on which child would continue education on the basis of grades. This seems like a very gender-neutral manner of deciding on children's education. However, given that boys and girls usually worked with the father or mother, respectively, in helping in household chores, and given that women's household chores were much more than those of men, it meant that the girls had less time to devote to their studies or other forms of self-development, like play. Consequently, even on the basis of grades attained, boys invariably continued to study longer than girls. The increased importance of men in cash earning activities and patrilocal marriage could only reinforce this bias in favor of boys' education.

New Policy Norms: Women's Inclusion in Forest Management

There is some hope for devolution, if governments and civil society institutions acknowledge the urgent need to close all spaces of the marginalization of indigenous peoples in general, and the indigenous women in particular, and provide them with effective assistance in reconstructing their present communities—based on gender equality and grassroots level self-determination in ownership and management of their forests, knowledge, sociocultural practices and political systems.

Women's groups as an important site for women's strength and mutual learning are increasingly accepted in national political and educational ideology. This is also seen in devolution policy and its implementation. In Asana village, Bastar, India, forest-based indigenous women have effectively been running Van Dhan Samitis (Forest Producers' Cooperative Societies), largely as a result of the women's struggle, but also facilitated by the space created by devolution. However, the Asana success also shows that unless women are organized, the space created by devolution would be usurped by local youth, who are not involved in forest production or protection.

In Orissa and Uttar Pradesh, the inclusion of women has usually meant that immediate needs like provision of wood fuel, and protection of water sources are considered in forest management decisions. But this may be achieved by creating separate "women's forests" within the community forest (see Box 2). In Nepal too, while the men-dominated community forestry schemes ignored the practical needs of fuel and fodder, the Nepal Leasehold Forestry, with women's user groups, did include their needs of fuel and fodder (Subba, 2000).

Box 2

In two villages in UP, women's groups had been able to carve out separate spheres for forest management by women. In Kail, the women used to articulate their frustration over the time and effort required for fetching daily requirements of fuelwood and fodder from the forest at a considerable distance from their village, especially if they to had to leave young children at home. The progressive-minded village head, recognizing the limited space available to women to influence decision-making in the male dominated village council, encouraged them to enclose a patch of degraded common land nearer the village, and start regenerating it. Initially this met with a lot of opposition from the men, particularly those who had to take care of domestic chores while the women patrolled their land. Some accused the village head of inciting women, while others voiced the usual opposition that the entire village had rights in the land.

Gradually the patrolling started yielding results and the women have begun to meet their fuelwood and fodder requirements from their own forest. This has since been registered as a separate "Mahila Ban" (Women's Forest). They rarely use the general village forest, as they have carved out their own separate resource management domain.

Simultaneously, there is also evidence that where women and men are in joint groups, men often marginalize the women, at least in the beginning. Village natural resource management groups in northeast India now do include women, but they are forced to remain silent, while the men speak and take decisions. One Naga woman explained why she is silent in these joint meetings, "If I say something different, my husband will beat me when we go home." (The authors' interview, 2001.)

Even without threats of beating, women are expected to be coy and shy, and not speak up in meetings. A woman from a forest-based community in the northern Vietnamese highlands passed a note to the meeting facilitator to request inclusion of very simple matters, like education for children and women in community decisions. Though there was nothing exceptional about these demands, she was still too afraid to speak in the meeting in the presence of the (male) village head.

Formalization of external relations and an increase in dealing with the bureaucracy disadvantage women. As Tiplut Nongbri points out in the context of the Indian Supreme Court's order on logging:

> The concept of the working plan mooted by the Supreme Court, according to which forests can be used only in accordance with centrally approved plans by the State Government ignores women's role in resource generation and also intensifies men's control over them ... experience shows that whenever women had to interact with the state machinery, they invariably fall back upon their brother, husband or son in executing their affairs (2000b: 12).

Nevertheless, inclusion of women in forest decision-making does make a difference in the administration of local management and also for gender relations. In the Bajawand Block Committee in Bastar, central India, with a membership of 3,000 covering 25 villages, the woman president, Kalavati, introduced a major change in the *tendu* leaf collection cards. Now, these cards have the name of women who are the primary collectors, rather than that of the male head of the household, as it used to be. As a result the membership of the cooperative has increased from just 10 percent women to 90 percent women. Women are better able to control the income for *tendu* leaf sales, and have more influence over the cooperative's decisions on sale conditions and *tendu* leaf management. The new norms, though still problematic in their functioning, do open opportunities for women to regain some of their lost control over forests, and over their own livelihoods.

Forest Relations and Knowledge

Women's knowledge of forests as a social resource can be useful for the effective management of forests. However, the inferior position of women in male-dominated hierarchy-based gender relations inhibits the use of this knowledge. It appears that there are two components to women's difficulty in contributing to social capital. First, the domestication of women is often reinforced or recreated through devolution of forest management. Second, women are subject to large-scale exclusion from forest decision-making.

While the state and market agencies are under pressure to involve women, they do so in many cases to appear participatory and gender sensitive. The women involved in NTFP collection, organized by forest producer cooperatives, have complained about the role of such officials and local NGOs too. In Bastar in central India, Nandini Sundar (2000) noted that where women have been protecting the forests, formalization into Suraksha Samitis (protection committees) has shifted power into the hands of the village, notably men. Likewise, in the case of Orissa, the youth clubs hardly ever took account of the women's knowledge and work in forests. The women have continued to work harder and longer under the worst conditions, with poor access to health care and education. How do the forest women build social capital—trust, cooperation and networks—in a situation where there is no social acknowledgement of their traditional, experiential knowledge of forests, and where they are denied a voice and participation in control and management of their forests and communities? Unless the power- and hierarchy-based gender relations are fundamentally transformed there can be no right kind and right amount of social capital for forest women.

Even within the community, spaces for women's participation in decision-making (households, communities and forest management committees) are plagued by male hierarchy and power-based gender relations. Many case studies of forest communities have reported that when women started protecting their forests and began managing the local forest resources, they faced problems with the men of their own and neighboring villages (He Zhonghua, 2001; Sarin, 2000; Sundar, 1997; Vasundhara, 2000).

The large-scale absence of women from formal local forest management is not because of women's backwardness in formal education; rather it is a manifestation of male dominance at all levels in decision-making, which translates at the local level into a general reluctance to involve women in local forest decision-making. As Harawati, leader of a forest producers' cooperative in Bastar, said, "You people (men) say what can women do, but if you don't tell them about the meetings, then really what can they do?" (Sundar, 2000: 24).

Harawati, like other women activists, has faced threats and even been thrown out of her caste for her work.

Indigenous women represent multiple traditions of knowledge of forest resources and technologies, but these have often been considered as non-knowledge because, as is the case with most such knowledge, it is represented in terms of being "sacred," and disguised in ritual and myth. There is, however, a growing recognition that indigenous peoples across the regions were able to develop forest management systems in a manner that was participatory and sustainable. Women had an important role to play in this management of food (both gathering and hunting), fuel, fodder, and cultivation of swidden fields. They had knowledge of seed varieties and storing techniques. Several studies also refer to women's hunting expeditions in the forests, and the annihilation of wildlife (Bosu Mullick, 2000; Kelkar and Nathan, 1991; Munshi, 2000; Porodong, 2001; Singh, 2000). It seems, however, that women's roles as hunters "have faded out of folk memory as gods as hunters emerged" (Sontheimer, 1997: 102). The anthropological approaches to hunter-gatherers in their gender specific roles: men as the hunters and women as the gatherers collapsed with feminist revisioning of hunter-gatherer research in the 1970s. There is, however, merit in the concept of an archaeological approach to hunter-gatherers; it is in the recognition that most of the world's hunter-gatherers are dead, that they "must be studied by excavation rather than interview, and that all this has some bearing on the kinds of theories in which students of hunter-gatherers ought to be interested" (Bettinger, 1987: 121).

The *bobolizans* in Sabah, for example, play a decisive role both in the selection of the site of the village as well as that of the swidden fields; no one in the community, including the village head can question her. Her role covered important matters such as the future of newlyweds, healing of the sick, fertility of the soil and safety of the village from a potential threat of supernatural powers. The *bobolizan* has good knowledge of animals, soil, water sources and forests, including *puru* (a protected forest, inhabited by spirits) and *himban* (a primary forest undisturbed by the human presence). Besides, to be a perfect *bobolizan*, she had to achieve two additional skills. One, expertize in rites and a great deal of memorizing to conduct daylong, even weeklong rituals without making mistakes; she has to acquire this knowledge through oral transmission from her mother or a close relative. Two, an expertize in the technology of weaving a very intricate design *surip* motif. A man could also become a *bobolizan*, but not of a higher order, since he would not concentrate on memorizing and would easily be distracted. Also, a woman could be a village head, but with the advent of modernity, Christianity and male dominance, they have begun to regard this position as suitable for a man.

The matrilineal and politically active Lua women of the 1978 revolutionary base of Nan province of Thailand described to Cholthira Satyawadhana (2001) the "myth" of Lua women's knowledge and technology of salt production. The grandmother, Yaa Lua, who discovered the salt mine and used salt in cooking, was watched by men of the community in order to discover her secret knowledge of salt. She was then executed by these men for her knowledge of salt. Many other forest communities around the world also have similar myths of knowledge stolen by men from women—the Mundurucu of the Amazon, various Papua New Guinea communities with their secret musical instruments, which women are not even supposed to hear etc. Myths are not history, but the struggles they represent have some correspondence to historical struggles over knowledge and control of technology.

Democratization of internal relations as a consequence of political struggle, however, is much less seen in the case of gender relations. Women are mobilized in political and forest control struggle and even organized for fulfilling various functions of support and direct participation too. But their representation in political councils, whether at the village or higher levels is usually nonexistent. They may be in the political organizations, though in very few numbers. The exclusion of women from traditional village councils of forest-dwelling communities, however, has not been seen to change in the course of the above struggles.

Further, to the extent there is a mobilization of women in the struggle there is usually demobilization after the struggle, as women go back to their usual functions. But the fact of their having been organized during the struggle does continue to have its effect. Whether the same organizations of women, or others that come up after the struggle, they have taken up issues related to women and gender relations. The most prominent of these have been the widespread movements of women's groups, both in the areas of Andhra Pradesh and in northeast India to ban the sale of factory-made liquor. Although articulated as a demand on the state, it is really an attempt to overcome what is a family matter of the use of income. The attempt, whether one agrees with the measure or not, is really to try to reduce men's overuse of family income for consumption of alcohol to the detriment of the welfare of the family, particularly children and women. This is one attempt at redefining gender relations that is common across these areas, though, of course, these movements are not confined to such areas.

The mobilization of women has made it easier for them to take up various livelihood programs in, say, IFAD projects in Andhra Pradesh and Manipur. In the course of functioning of these groups, it has been observed, as is common with microcredit groups of Grameen Bank and other NGOs, that women's groups are more reliable in terms of maintaining repayment schedules, which enhance group solidarity. In the tree plantation and nursery schemes in Nagaland, it is reported that women-managed nurseries and plantations do better than those managed by men. The norms of reciprocity and trust seem stronger among women than among men.

At the same time, it has been difficult for women to effectively enter formal forest management groups. The exclusion of women from traditional village councils is, in a sense, carried over into formal JFM committees. Where rules insist on the inclusion of women, this often remains formal. But the formal inclusion of women in these management committees raises the possibility for women's organizations to struggle to make this inclusion real.

It is also noticed that separate women's groups tend to develop women's management capabilities more than mixed groups. In mixed groups there is a tendency for men to dominate the important functions, while in women-only groups, women get more chances to develop their managerial skills, as they are responsible for all the aspects of the group's functioning. What this means is that we have to take account of gender relations in developing norms of reciprocity and trust. Women's groups seem to have better group solidarity than men's groups; and separate women's groups are needed to develop women's capabilities.

Political Struggle for Women's Resource Management

There is a type of political struggle that can be identified as not only sufficient for developing the capability to manage forests and other resources efficiently and effectively, but also necessary for this purpose. This is the internal political struggle within the group or community concerned.

To state the proposition in a general manner: In any situation, there are always those who benefit from the existing norms or from the laissez-faire situation following the breakdown of those norms. Those who benefit from the existing norms (or lack of norms as the case may be) will oppose the setting up of new norms. This resistance will need to be overcome by struggle/mobilization of those who are likely to benefit or expect that they will benefit from the new norms. If the new norms are necessary for building the capability of the community to deal with its internal forest management problems, then we can say that political struggle is necessary for setting up new norms, or in moving from a laissez-faire situation to one of accepted norms as the case may be.

Let us take the situation in the Khasi areas of Meghalaya. When the village *sardars* (headmen) or other notable village men and their families were grabbing community forests and registering them as their own, or more often their wives' private lands; the poorer sections of the community, who were being left out of this process, mobilized themselves and forced a change in the rule regarding community forests—the community forests were then divided up more or less equally among all families of the village community. (Since the Khasi are matrilineal, women own the ancestral land, though this is fast changing.) This village-level mobilization instituted new forms of ownership and control of former community forests, which led to more care of the forests, management practices of various kinds and even enrichment planting, all of which increased the community's capability to efficiently manage its forests.

A more telling and frequently occurring example of internal political struggle relates to the inclusion of women in forest management groups. It has already been established earlier that the exclusion of women leads to inefficiency in community forest management, since their exclusion means their local knowledge is not adequately taken into account, and that the requirements of the areas of livelihood that are women's responsibilities (fuelwood most eminently) are ignored in management decisions taken by men alone. Whether in Haryana, Andhra Pradesh or northeastern India, effective inclusion of women in community forest management has required mobilization by the women, and their overcoming of resistance from men. A feature of this political struggle has been that external agencies, project management offices, forest department officials, NGOs, and women's movements in particular, have all played facilitating roles in this internal political struggle. In the notable example of the International Labor Organization–Center for Women's Development Studies (ILO–CWDS) Bankura afforestation project, the project director went so far as to threaten that there would be no project if the men did not transfer the title to the degraded land to their wives.

Whether formally or informally, openly or covertly, establishing new norms requires political struggle. Forest-based communities have gone through, and in some cases, are still going through, protracted struggles to establish patriarchal norms. As analyzed in detail elsewhere (Kelkar and Nathan, 1991; Nathan, Kelkar and Yu Xiaogang, 1999) the demonization of women, often in the form of persecution for practicing malevolent witchcraft, is a form of the struggle through which men excluded women from the spiritual and material management of community affairs. Simultaneously, and sometimes often in the same communities, there are also struggles to overcome these patriarchal norms.

Thus "taking account" of gender relations is not something that develops by itself. The inclusion of women in management committees also results from political struggle of various kinds. The difference in these gender struggles is often that external agencies too have a substantial role to play. For instance, external project rules requiring the inclusion of women in committees, or legislation on the same matter, are important factors in bringing about changes in gender relations. What is necessary to note is the dialectical relations between internal struggles and external enabling rules and decisions. Each one feeds into the other.

NOTE

1. The 12 studies and the research team include: the matrilineal Mosuo of Yunnan, China (by He Zhonghua); the matrilineal Rungus of Sabah, Malaysia (by Paul Porodong); the Tamangs of Nepal (by Suman Subba); the Khasi Matriliny in Meghalaya, India (by Tiplut Nongbri); Women in the West Khasi Hills (by Patricia Mukhim); the matrilineal Lua in Thailand (by Cholthira Satyawadhana); Management of the Forest Resources in Lijiang, China (by Yang Fuquan); Women and Forest Management in Lashih Lake, Yunnan, China (by Xi Yuhua); Women and Forests: The Warlis of Western India (by Indra Munshi); Gender Relations and Witches among the Indigenous Communities of Jharkhand, India (by Samar Bosu

Mullick); Gender Relations among the Mizos, Northeast India (by Linda Chhakchhuak); and Women as Hunters in History (by K.S. Singh). These studies were conducted as part of the IFAD sponsored research project: Creating Space for Local Forest Management in Asia, jointly coordinated by CIFOR and AIT. We would like to thank Phrang Roy and Ganesh Thapa of IFAD for having taken the initiative for sponsoring these 12 gender relations studies. We thank David Edmunds and Eva Wollenberg of CIFOR for their painstaking comments on this paper. We appreciate and acknowledge the inputs of Madhu Sarin and Neera Singh.

REFERENCES

Bachofen, Johann J. (1861) The Mother Right, Das Mutterrecht, Stuttgart.

Bettinger, Robert L. (1987) 'Archaeological Approaches to Hunter–Gatherers,' in *Annual Reviews of Anthropology*, 16: 121–42.

Bosu Mullick, Samar (2000) 'Gender Relations and Witches among the Indigenous Communities of Jharkhand, India,' *Gender, Technology and Development*, 4(3), pp. 333–58.

Chhakchhuak, Linda (2000) 'Gender Relations among the Mizos,' unpublished.

Cholthira Satyawadhana (2001) 'Appropriation of Women's Indigenous Knowledge: The Case of the Matrilineal Lua in Northern Thailand,' *Gender, Technology and Development*, 5(1), pp. 91–111.

Divale, William (1984) 'Matrilocal Residence in Pre-literate Society,' Monograph, UMI Research Press, Ann Aarbor.

Guo Da Lie (2000) 'The History of the Ecology and Forestry Administration in Naxi District,' unpublished.

Gough, Kathleen E. (1959) 'The Nayars and the Definition of Marriage,' *Journal of the Royal Anthropological Institute*, 89: 23–34.

He Zhonghua (2001) 'Forest Management in Mosuo Matrilineal Society, Yunnan, China,' *Gender, Technology and Development*, 5(1), pp. 33–62.

Kelkar, Govind and Dev Nathan (1991) *Gender and Tribe: Women, Land and Forests in Jharkhand*, Kali for Women, New Delhi and Zed Press, London.

Mann, K. 'Is Matriliny a Symbol of Higher Status? A Case of Garo Woman.' *Man in India*, March 1987.

Mukhim, Patricia (1999) *The Shillong Times*, Feb. 19.

Munshi, Indra (1986) 'Tribal Women in the Warli Revolt 1945–47: Class and Gender in the Left Perspective,' *Economic and Political Weekly*, 21(17), April 26.

——— (2000) 'Women and Forest: A Study of the Warlis of Western India,' unpublished.

Nathan, Dev, Govind Kelkar and Yu Xiaogang (1999) 'Demonization of Women: Cross-Cultural Analysis of Naxi, Dai and Santhal,' *Naxi Religion, Gender and Culture*, Gender Studies Monograph 8, Asian Institute of Technology, Bangkok.

Nongbri, Tiplut (2000a) 'Khasi Women and Matriliny: Transformations in Gender Relations,' *Gender, Technology and Development*, 4(3), pp. 359–95.

Nongbri, Tiplut (2000b) 'Timber Ban in Northeast India: Effects on Livelihood and Gender,' unpublished.

Porodong, Paul (2001) 'Bobolizan, Forests and Gender Relations in Sabah, Malaysia,' Gender, Technology and Development, 5(1), pp. 63–90.

Richards, A.I. (1950) 'Some Type of Family Structure among the Central Bantu,' in A.R. Radcliffe-Brown and Daryll Forde (eds.), African Systems of Kinship and Marriage, Oxford University Press, London, pp. 207–51.

Sarin, Madhu (1996) 'Who is Gaining? Who is Losing? Gender and Equity Concerns in Joint Forest Management,' Working paper by the Gender and Equity Sub-group, National Support Group for JFM, Society for Promotion of Wastelands Development, New Delhi.

——— (2000) 'India Country Chapter,' in Antonio Contreras, Liu Dachang, David Edmunds, Govind Kelkar, Dev Nathan, Madhu Sarin, Neera Singh and Eva Wollenberg (eds.), Creating Space for Local Forest Management, CIFOR, Bogor, unpublished manuscript.

Schlegel, Alice (1972) Male Dominance and Female Autonomy: Domestic Authority in Matrilineal Societies, HRAF Press.

Scott, James (1985) Weapons of the Weak: Everyday Forms of Peasant Resistance, Yale University Press, New Haven, USA.

——— (1990) Domination and the Arts of Resistance: Hidden Transcripts, Yale University Press, New Haven, USA.

Singh, K.S. (1999) 'Changing Attitude to Forest and Nature: A Historical Review with Focus on Jharkhand,' unpublished.

——— (2000) 'Shamanism, Witchcraft and the Position of Women in Tribal Society,' unpublished.

Singh, Neera (2000) 'Forest and Communities in Orissa,' in Antonio Contreras, Liu Dachang, David Edmunds, Govind Kelkar, Dev Nathan, Madhu Sarin, Neera Singh and Eva Wollenberg (eds.), Creating Space for Local Forest Management, CIFOR, Bogor, unpublished manuscript.

Shih, Chuan-Kang (1993) 'The Yongning Moso: Sexual Union, Household Organization, Gender and Ethnicity in a Matrilineal Duolocal Society in Southwest China,' Ph.D. Dissertation, Stanford University, unpublished.

Sontheimer, Gunter-Dietz (1997) King of Hunters, Warriors and Shepherds: Essays on Khandoba, edited by Anne Feldhaus, Aditya Malik and Heidrun Bruckner, Manohar Publishers, New Delhi.

Subba, Suman (2000) 'Transition of Gender Relations in Tamang of Nepal,' Kathmandu, Nepal, unpublished.

Sundar, Nandini (1997) Subalterns and Sovereigns: An Anthropological History of Bastar, 1854–1996, Oxford University Press, Oxford.

——— (2000) 'Critical Analysis of Involution in Bastar,' A study done for CIFOR, Bogor, unpublished.

Townsend, Janet G. (1995) Women's Voices from the Rainforest (International Studies of Women and Place), Routledge, London.

Vasundhara (2000) 'Women's Concerns in Community Forest Rights Debate in Orissa,' in Antonio Contreras, Liu Dachang, David Edmunds, Govind Kelkar, Dev Nathan, Madhu Sarin, Neera Singh and Eva Wollenberg (eds.), Creating Space for Local Forest Management, CIFOR, Bogor, unpublished manuscript.

Wickramasinghe, Anoja (1999) *Gender Aspects of Woodfuel Flows in Sri Lanka: A Case Study in Kandy District*, FAO Regional Wood Energy Development Programme in Asia, Bangkok.

Xi Yuhua (2000) 'Forest Inspection of EnZong Village, Lijiang, Yunnan,' China, unpublished.

Yang Fuquan (2000) 'The Investigation of Use and Management of the Forest Resources in Longquan Administrative Village of Baisha Township of Lijiang County, Yunnan, China,' Kunming, China, unpublished.

I

History and Myth

Gender Roles in History: Women as Hunters

K.S. Singh

The critique of the anthropological view of "Man the Hunter" has focused on the role of women as gatherers—gathering being more important than hunting in many phases and situations—and as the mainstay of the family with the bonding of infant and mother (Zilman, 1998). Hunting is said to have emerged out of gathering at a later stage. While men went on to become the primary hunters of wild animals, particularly major ones, women continued to gather, fish, and even hunt minor animals. Men acquired the knowledge to use heavy weapons for killing major animals, while women used light weapons. This conventional wisdom of gender roles is being questioned further not only by feminist anthropologists, but also by archaeologists, historians, folklorists, etc. The objective of this chapter is to present evidence from India on the subject in terms of archaeological, folkloric and historical evidence, and also contemporary anthropological material.

Scholars of rock art paintings generally say that it is difficult to identify genders in the earliest paintings showing hunting, but these do exist. The Neolithic site at Burzhom in Kashmir shows a woman and a man hunting a big animal with a big spear, demonstrating that women had a major role in hunting (as illustrated on the cover). Current literature speaks of hunting as a collective effort and the participation of women as beaters and as transporters of game among the San, the hunters and gatherers of Kalahari (Tanaka, 1980: 165). However, the following evidence will prove that women have been hunters and that their role as beaters today is only a pale reflection of this earlier status.

The ancient texts including the Greek accounts and Kautilya's *Arthshastra* have many images of women as soldiers and hunters. Female bodyguards accompanied the Mauryan king on hunt, riding on chariots, horses and elephants, carrying all kinds of weapons. Women-archers would surround a king when he would rise from his bed. One of the Bharhut sculptures show a woman riding a horse, fully caparisoned and carrying a standard. The image of a woman hunter (*svaghni*) cutting flying birds into pieces is said to be as old as the *Rigveda*. A 12th-century sculpture in Rajasthan shows a woman hunter carrying a stick with animals hanging from each side of the ends, while a man stands on the left handing her arrows (Jaiswal, 2001). These historical images are reflected well in the background of the *Puranic* themes describing women as riders of animals and their killers.

India probably did not have a counterpart of the Greek and Roman hunting goddesses, Diana and Artemis (Field, 1977.) Shiva was the hunter. There also existed folk gods who were hunters, pastoralists, shepherds, and small shrines of such divinities (probably female) daubed with vermilion, dotting the forest regions or fringes where they were propitiated by hunters, travelers, and now by contractors and members of the timber mafia, who have even put up permanent structures in such places. Khandoba of western India is a folk god who came to be identified with Shiva. Shiva as *Sabara* or hunter saved Arjuna by killing the demon Malla, and is hence known as Mallari in western Indian mythology. The interesting point here is that a minor hunter, Goddess Malachi, accompanied Shiva on this expedition (Sontheimer, 1997: 102, 212). Shiva as a hunter in the form of *Kirat* (Mongolian) humbled Arjuna when they quarreled over the hunting down of a wild boar. It seems that the roles of minor female divinities as hunters have faded out of folk memory as gods emerged as hunters.

Folklore and *Puranic* myths glorify Mother Goddesses as killers of major wild animals and we still see depictions of Mother Goddesses riding wild animals. For example, the Mother Goddess cult in eastern India, particularly Bengal, which has assumed various forms of gorgeous iconography, shows the Mother Goddess killing a wild buffalo or a demon in the form of a wild buffalo and riding a lion, instead of tiger, which is unique to the ecosystem of the region. The lion motif in this iconography and textual material seems to have traveled from northwest India, where lions existed till the close of the precolonial period, to eastern India. The Goddess is also called *Sinh-vahini* or lion-rider and *Mahisasur-mardini* or killer of the demon in the form of a wild buffalo. This suggests that women taming or killing wild animals was not considered strange.

While the ancient texts make specific references to hunting by women, there were instances where women also accompanied their spouses to the deep forests (e.g., Pandu and his wives, Draupadi and her husbands) or to the battlefield (e.g., Dasarath and Kaikeyi). The *Dharmashastras* (Books of Laws) which generally deal with the norms of a settled agricultural society neither permit nor prohibit hunting by women. There was no strict taboo imposed on women with regard to hunting, use of weapons and shooting major wild animals in Hinduism or Islam, or by custom as is borne out by two well-known examples.

Rani Durgavati, a Chandel princess married to a Gond prince of Garh Mandla, the capital of the Gond Kingdom, a contemporary of Akbar, whose exploits have been described in the *Akbarnama*:

She 'was a good shot with gun and arrow, and continually went hunting and shot animals of the chase with her gun. It was her custom that whenever she heard that a tiger had made his appearance, she did not drink water till she had shot him. There are stories current in Hindustan of her feasts and her frays' (Abu-l-Fazl, n.d.: 2:327).

Mughal queen Nur Jahan is reported to have trained herself as a markswoman. Her fond husband, Jehangir, records that in the 12th year of his reign:

On the seventh as the huntsmen had marked four tigers, when two watches and three *gharis* had passed, I went out to hunt with my ladies. When the tigers came in sight, Nur Jahan Begum submitted that if I would give her order she herself would kill the tigers with her gun. I said, "Let it be so." She shot two tigers with one shot each and knocked over the two others with four shots. In the twinkling of an eye she deprived of life the bodies of these four tigers. Until now such shooting was never seen, that from the top of an elephant and in sight of a *howdah* (*amari*) six shots should be made and not one miss, so that the four beasts had no opportunity to spring or move. As a reward for this good shooting I gave her a pair of bracelets (*pahunchi*) of diamonds worth 100,000 rupees and scattered 1,000 as *ashrafis* (over her), (Beveridge, 1968: 1:371).

While this shooting exploit took place in Malwa region, the other incident occurred in Mathura where Jehangir camped. He narrates:

The huntsmen reported that there was in that neighborhood a tiger that greatly troubled and injured the ryots [serfs] and wayfarers. I immediately ordered them to bring together a number of elephants and surround the forest and at the end of the day myself rode out with my ladies. As I had vowed that I would not injure any living being with my own hands, I told Nur Jahan to shoot at him. An elephant is not at ease when it smells a tiger and is continually in movement and to kill with a gun from a litter (*imari*) is a very difficult matter in as much as even Mirza Rustom, who after me is unequalled in shooting, has several times missed three or four shots from an elephant. Yet Nur Jahan so hit the tiger with one shot that it was immediately killed (ibid.: 2:104–5).

The two incidents illustrate, as mentioned earlier, the use of sophisticated arms by women to kill tigers. They also underline the fact that the ladies accompanied the emperor on hunting expeditions. Hunting expeditions in general have been described in both Sanskrit and Persian texts. There are other examples too. Shahjahan's queen, Mumtaz Mahal, shot minor animals like black bucks (*ahu*). The Rajput girls would be exhorted to shoot and kill animals. In the early 13th century, Razia Sultan would dress like a man and go out hunting (personal communication, Ranjit Singh). The above instances also illustrate that the state in historic times would intervene in rural India to control vermin. Mahesh Rangarajan says (1999a: 59):

The Indian legacy of dealing with carnivorous animals was rich and diverse. In contrast to Britain and Western Europe, there was no long history of state-sponsored projects to eliminate carnivores. Rulers did kill man-eaters and help in the process of agrarian expansion. But a degree of tolerance was perhaps easier in the context of relatively dispersed rural settlements with a large proportion of the land area under forest cover. There is evidence of recourse to religious and magical methods to ward off tigers. But there were also a variety of other means of self-defense ranging from avoidance to simple elimination of marauding animals.

The reason for giving this long quotation is to contrast the precolonial with the colonial situation that followed, when annihilation of wildlife was practiced both literally and metaphorically. Hunting was practiced with a vengeance as a manly pursuit, till a later phase when elements of a conservation policy emerged. As Mahesh Rangarajan tells us (1999b: 131):

> The world of *shikar* was pre-eminently a man's world. Hunting anecdotes are replete with references to 'manly pursuits', to pig-sticking as a means for 'the purgation of lusts' and to the need to hunt to overcome 'effeminacy'. British writers often claimed that only white men and a few Indian communities like the martial races (Sikhs, Gurkhas) and forest tribals (Toda, Santhal) were brave enough to hunt dangerous game on foot.

That hunting was thus necessary to stress man's prowess, and his status is a recurring theme of both colonial and precolonial literatures, which served to obfuscate the role of women as hunters. But even in the colonial period there is a story of Isabel Savory, who killed Himalayan black bears and wrote a book, *A Sportswoman in India: Personal Adventures and Experiences in Known and Unknown India* in 1900, to challenge men's monopoly in the field. She, "shared a big game hunter's view of India: she disliked Indian trappers and fowlers, even as she extolled the glories of European sport" (ibid.: 131–37). She had all the usual colonial biases, but she was not alone in her pursuit, even though women's participation in hunting was not well documented.

In parts of Bengal, for example, the statistics on wild animals killed in Santhal Parganas during the period 1858–63 showed that leopards, hyenas, and bear came first, followed by tigers and wolves (Sivaramakrishnan, 1999). Towards the close of the 19th century as elements of a conservation policy emerged, the vermin extermination operations slowed down. There was a softening of attitude towards tigers as they came to be seen as an embodiment of gentlemanly virtue, and their positive role as predators was recognized (Rangarajan 1999a: 65). However, the status value attached to a tiger killer continued to be universally recognized as the maharajahs among others, and scions of their families, including some women, went on adding tigers to their trophies. The maharanis of Dhar in Malwa and Indore are noteworthy examples. Gayatri Devi shot a panther at an early age. She recalls:

> Early in the morning, the news was brought by some villagers that a panther had to be killed in a nearby area, so after lunch Indrajit, Menaka and I set off. Each of us was mounted on a howdah elephant with an ADC behind us, and Indrajit was specifically instructed to let me have the first shot. Naturally, we had all been taught from the time we were very young how to shoot, how to be careful, how to make sure that we got a clear shot without the chance of wounding any of the hunting elephants through overexcitement, and so on. We were in a very small jungle that afternoon, near a village, and we could hear the elephants trumpeting as they usually did when there was a wild or dangerous animal in the vicinity. Then the breathless moment came when the beat began.
>
> By the standards of more experienced hunters, my first triumph might well seem rather tame. When the panther was finally forced out of cover, it snarled once and simply stood still, staring at my elephant. The ADC behind me told me to fire and the only thing I can say to my credit is that I didn't lose my nerve. I picked up my gun—I used a twenty-bore shotgun—and got him in the face with my first shot. There was great joy and jubilation; even the mahouts and the professional hunters with us joined in. I was deluged with congratulations, and when we got back to the palace everybody made a big fuss over me. We sent a telegram to Ma in Delhi telling her that I had shot my first panther. I was twelve years old and speechless with pride and excitement (Gayatri Devi and Shanta Rao, 1976: 65–66).

The postcolonial situation saw, literally, the opening of a floodgate, with the depletion of forest resources and merciless killing of wildlife, particularly major mammal populations. There are instances in the family when women would train themselves as shots and also *participated in killing tigers.* There are umpteen examples of women having killed a major animal when it posed a threat to the life and property of the family members. In our family, for instance, two women sought tigers after a rigorous training in shooting wild animals including tigers in the 1940–50s. Likewise, a couple of queens in the princely states of Jaipur and Rajgadh, in colonial and post-colonial periods in western and northern India, sought and hunted tigers. Instances of such elite women hunters apart, even ordinary women, especially among forest-dwelling societies, have hunted big animals that have threatened their lives and property.

Prohibitions for Women

We now turn to the contemporary anthropological situation. A few general comments will be in order. First, while there are a whole range of prohibitions for women on activities ranging from weaving a cot to ploughing or roofing, or sharing certain portions of sacrificial meat, there is no specific prohibition regarding hunting or allied activities, such as handling weapons. In fact, literature on the Chota Nagpur tribes is specific about the absence of any prohibition against women touching weapons or using them or dancing with them. There is, however, a general prohibition about anybody, including a woman, stepping over a weapon in all communities. Similarly, menstruating women are prohibited from participating in any activity leading to a hunting expedition. Not only the menstruating woman, but even her male relations cannot participate in a hunting expedition (Rosner, 1982: 32). Pregnancy is considered the most detrimental factor for the success of the hunt. As Rosner says:

> Not only is no contribution asked for from a family where a woman is carrying a child, but such a woman is not allowed to come out of her hut on the day of the hunt and her husband is not permitted to take a share of the head of any animal killed that day (ibid.: 34–35).

A married woman is expected to remain chaste during a hunting expedition to ensure the success of her husband in the hunt. A woman with a broken pitcher standing in the path of the expedition is considered a bad omen. In other communities, a woman standing with a pitcher full of water on her head is considered an auspicious sign for a person proceeding on a journey or any expedition.

There are some major tribal hunts called *akhand sikar* in Baripada (Orissa) and *paradh* in Bastar from which tribal women are kept out, though the women, particularly the wives of the male hunters who die during the hunt, get a share of the meat of the animals killed. A reference has been made to the heavy weapons that men used to kill major animals. The Indian evidence suggests that the equipment used to kill major wild animals are light and the techniques used to trap them are simple (Rosner, 1982: 39–42), which presumably do not stand in the way of a woman intending to use them or being capable of using them. In Chota Nagpur these weapons consist of bows and iron arrowheads, trigger cord attached to an iron spike, etc.

There is also a whole range of totems connected with animals, small and big, that prohibit an indigenous or other person from killing or harming the totemic animals. Such totems stress the close relationships that exist between various forms of life and environment from the early days of human society. Closely related to this is the widespread tiger myth, which is common to the tiger areas of Southeast and South Asia. A tiger's soul can be transferred to humans or human beings can themselves be transformed into tigers. The interesting thing to note in these tiger myths is that the tiger soul can be transferred to women or women can be transformed into tigresses. Thus, women too can possess the tiger spirit. Verrier Elwin (1954) narrates a number of stories where a Gadaba woman gave birth to a tiger cub, the clay model of the tiger became alive, a woman dreamt of a tiger and conceived, and five tigers lived on the branches of a great *buja* tree that grew out of the blood of a woman killed by a shaman. Reports about women possessing the tiger spirits in Khasi society as well as in central India also suggest that the notion of women accepting the tiger spirit was normal. There was no gender specific discrimination (this is seen in Malaysia also). Given these facts, there is no prohibition against women going into deep forests, which is generally described as mysterious. In fact, women as gatherers have been scouring deep forests for ages.

Jani Shikar

This brings us to the mythology of the unique ritual hunt practiced by indigenous women, mainly belonging to the Oraons, popularly known as *jani shikar* (women's hunt).

Baro Bachare Raja,
Jani shikar vaini ka munde raja pagri bandhai
(O Raja, every 12 years, we go out on a women's hunt wearing a
turban on the head).

Singing this couplet, the women dress like men, complete with
turban, and hunt wild animals. This ritual hunt is still alive in India
as it has been during the 40 years of my working life (1959–99). The
roots of this phenomenon go deeper than a mythical/historical story
of its origin. On the surface, this ritual is traced to a historical incident, when the Oraons, who claim to have been in possession of
Rohtas fort in Kaimur hills, lost control of it as a result of a stratagem resorted to by the enemies and with it, their preeminent
position. At the point when the menfolk were drunk, the women
came forward to fight the battle, but they lost. The earliest version
of this story is:

> The Oraons say they once dwelt on the Rohtas plateau under a
> Raja or king of their own tribe. The place was well fortified so as
> to defy the strongest enemy. The Oraons had erected a stone
> rampart about a mile in height, and the enemy long sought in
> vain to effect a breach. At length the 'Hakims' caught hold of a
> milkwoman of the Ahir caste who used to supply milk to the
> Oraon Raja, and who had therefore free access to the fort.
> Inducements were offered to this woman to suggest to the
> enemy a practicable means of occupying the fort. She accord-
> ingly advised them to wait till the morning of the ensuing Khaddi
> or Sarhul festival when all the Oraon males were sure to get
> dead drunk. This turned out to be correct, and the enemy fol-
> lowed her instructions and succeeded in entering the fort.
> Although the Oraon women who had been pounding rice with
> their wooden pestles to prepare bread for the Khaddi festival,
> came out with their pestles and valiantly met their foe, these
> amazons were soon overpowered. The Oraon Raja and his sub-
> jects, it is said, fled the fort through a subterranean passage
> known only to themselves. The enemy lighted huge torches
> "each of which consumed four maunds of oil", but they failed to
> discover their exit (Roy 1915: 29–30).

The stratagem referred to here can be located in history in the course
of the capture of Rohtas fort by Sher Shah. As Abu-l-Fazl (n.d.:
1:334–35) tells us:

Sher Khan is said to have avoided direct fight with Humayun and went off by way of Jharkhand to Bengal. When Bengal fell to Humayun in 1538 AD Sher Khan, having taken the choicest treasures of Bengal, went off by Jharkhand towards Rohtas and got possession of it by means of stratagem. It was held by a Brahmin, Raja Cintaman. Sher Khan begged him to receive his family, prepared six hundred litters and placed in each two armed youths, while maid servants were placed on every side of the litters. By this stratagem he took the fort, a very strong fort and blocked the road to Bengal.

It will appear from these two accounts that the origin of *jani shikar* had little to do with the capture of Rohtas. The Oraons never had a kingdom. The Oraon women were brave, and they did not go out to fight but to hunt. In fact, *jani shikar* has its origin in the long forgotten tradition of women being hunters. There was no doubt an acceptance of men's dominant role in hunting when women dressed as men and also took out their weapons to hunt. Further, they hunted small animals. It is a remnant of a hunting tradition in which women not only participated but also led the hunting; a role they gave up later; a grudging acceptance of the male role. In the song, wearing a *pagri* (turban) and dressing like a man suggests that there were occasions when they led hunting expeditions like men.

Rituals (like *jani shikar*) are increasingly being confined to nonreserved forest areas, as ideas of conservation of forests and wildlife come into acceptance by indigenous people, including women. They are, therefore, satisfied with hunting of domesticated animals like pigs and domesticated birds like fowls. This does not mean, however, that the ideas of conservation are universally accepted. There are still instances where women go out hunting in forests where they are prevented by forest guards. Current accounts of *jani shikar* suggests that with the depletion of forest and dwindling of wildlife, the women move from village to village hunting down domesticated animals and fowls, or minor wildlife in areas where the forests survive. Moreover, this is done under the strict supervision of the forest department at a few places in order to avoid injury to wildlife or damage to the environment.

There is also a parallel ritual hunt called *desua sendera* (hunt of the country) practiced by the Hos, both women and men on certain occasions. To us, it appears that these ritual hunts are the surviving portions of a strong tradition among the Oraons and Munda indigenous people, reflecting the historical memory of women's role in hunting which was diluted and eroded over time. It is not just a case of role inversion, which allows for a measure of rupture from tradition in order to reinforce the denial of women's hunting rights. Without some continuing familiarity with weapons and their use, it would not be possible for women to suddenly take up these roles on ritual occasions.

To conclude, it should be noted that women's role in hunting in some situations was never specifically denied according to the evidence we have discussed in this article nor was it prohibited by custom or religious injunctions. Their role was no doubt restricted, and hunting practiced occasionally in the circumstances that existed. This finding should contribute towards an understanding of the role of women in an area where fierce debates have raged between the protagonists of "man the hunter" and feminist anthropologists. There was certainly an area, or a small gray area, where women had a role in hunting and where the taboos against women hunting did not exist or were weak.

In fact, it seems women did play a role in hunting in some hunting-gathering situations, a role which still prevails. It was only at the stage of agriculture that gender roles were probably clearly defined and hunting for women survives as a mere ritual. However, as mentioned earlier, there was never any prohibition, directly or indirectly expressed or enforced against hunting by women in some situations.

REFERENCES

Abu-l-Fazl (n.d. [reprint 1972]) *The Akbarnama of Abu-l-Fazl*, Vols 1 and 2, translated by H. Beveridge, *Bibliotheca Indica: A Collection of Oriental Works*, published by the Asiatic Society of Bengal, Rare Books, Calcutta.

Beveridge, Henry, ed. (1968) *The Tuzuk-I-Jehangiri, or Memoirs of Jehangir*, Vols 1 and 2, translated by Alexander Rodgers, Munshiram Manoharlal, New Delhi.

Elwin, Verrier (1954) *Tribal Myths of Orissa*, Oxford University Press, Calcutta.

Field, D.M. (1977) *Greek and Roman Mythology*, Chartwell Books, New Jersey.

Gayatri Devi and Shanta Rao (1976) *A Princess Remembers: The Memoirs of the Maharani of Jaipur*, J.B. Lippincott Company, Philadelphia and New York.

Jaiswal, Suvira (2001) 'Female Images in the Arthasastra of Kautilya,' *Social Scientist*, 29(3–4), March–April, pp. 51–59.

Rangarajan, Mahesh (1999a) 'The Raj and the Natural World: The War Against "Dangerous Beasts" in Colonial India,' Research in Progress Papers, Nehru

Memorial Museum and Library, New Delhi. Later published as 'The Raj and Natural World: The Campaign Against Dangerous Beasts in Colonial India,' *Studies in History*, 16(2), pp. 266–99.

——— (ed.) (1999b) *The Oxford Anthology of Indian Wildlife*, Vol. 1: *Hunting and Shooting*, Oxford University Press, New Delhi.

Rosner, Victor (1982) *The Flying Horse of Dharmes*, Satya Bharati Publications, Ranchi.

Roy, S.C. (1915) *The Oraons of Chotanagpur*, published by the author, Ranchi.

Sivaramakrishnan, K., (1999) *Modern Forests, State-making and Environmental Change in Colonial Eastern India*, Stanford University Press, California.

Sontheimer, Gunter-Dietz (1997) *King of Hunters, Warriors and Shepherds: Essays on Khandoba*, edited by Anne Feldhaus, Aditya Malik and Heidrun Bruckner, Manohar Publishers, New Delhi.

Tanaka, Jiro (1980) 'Discussions,' in *The San, Hunter-Gatherers of the Kalahari: A Study in Ecological Anthropology*, translated by David W. Hughes, University of Tokyo Press, Japan.

Zilman, Adrienne L. (1998) 'Woman the Gatherer: The Role of Women in Early Hominid Evolution,' in Kelley Hays-Gilpin and David S. Whitley (eds.), *Reader in Gender Archaeology*, Routledge, London and New York, pp. 91–105.

The Fireplace: Gender and Culture Among Yunnan Nationalities

YANG FUQUAN

At dusk, one can observe Aini men surrounding a fireplace sipping tea and smoking cigarettes or pipes, or gossiping about the details of their hunting experience or talking about the yield of crops in the fields. When the host greets you and asks you to be seated, you will also hear laughter from the other side of a bamboo fence partition. You will see the flashing fire through the bamboo fence, and occasionally smell the fragrance of cooking rice—over there is a fireplace around which are gathered only women.

Leaving the Aini village, we come to the Wa village in Menglian county. Climbing up to a big bamboo building, we also find there are two fireplaces in the building separating the women from the men. This phenomenon of the fireplace symbolizes gender division. In this case, fireplaces are classified into female and male fireplaces, each with its own name. For instance the female fireplace is called *Bialeng* whereas the male is *Bialai*. The fireplace signaling gender is correlated with gender hierarchies also reflected in wedding and household customs and protocols. For many nationalities, women do not possess legal membership in the fireplace, but in dual fireplace

nationalities women have more power within the family and society.
Wa women in Menglian, for instance, possess higher social status than
men, and the daughters instead of the sons inherit family possessions.

The Fireplace

As society developed, the fireplace accompanied humankind in the
association of the fireplace with the spiritual realm and with material
life. Many nationalities believe that the deity in the fireplace is
responsible for the sustenance of the family, it makes the family rich
in food and clothes, and is also important for their bare survival.
During festivals, people not only worship the fireplace and the god of
the fireplace, but also use the fireplace as a fortuneteller to foresee
harvests or famine in the coming year. People also associate the fireplace with their birth and ancestry. Women give birth around the fire
and it is believed that the fireplace is the incarnation of dead parents;
that the soul of parents resides in the fireplace.

The ritual of placing and lighting a new fireplace requires a new
house, and someone who is highly respected must be asked to light
the fire. The location of the fireplace in a new house is also selected
only after great consideration. It is still popularly believed among the
Lisu, for instance, that an intermittent noise will be emitted if a fireplace is set in an inappropriate place, and the noise will stop only if
the fireplace is moved to its proper place.

A number of taboos and kinds of worship of the fireplace have
been derived from the fear of the god of the fireplace. The taboos
regarding the fireplace range from prohibiting warming one's shoes
and feet at the fireplace to not trampling on the iron tripod, and in
some cases, to pregnant women not being allowed to sit beside the
fire. The Dai people believe that the three legs of the fireplace represent the three emeralds symbolizing well-being, peace and happiness. The Yi people regard the three stones of the fireplace as
ancestry, male and female. The god of the fireplace may descend and
bless people with peace and well-being or alternately bring down
punishment upon them. People not only worship the fireplace and
pray for its blessing, but also fear it.

Fireplace worship is found among almost all nationalities who have
fireplaces. They worship the fireplace during festivals, in the harvest
season, in times of sickness and when giving birth. The Lisu and Yi
people slaughter hogs and fowl during their national celebrations

and worship the three pot-posting stones in appreciation to the god of the fireplace. The Mosuo people worship the fireplace for blessings from the souls of their dead parents by contributing food to the god of the fireplace in their daily meals thrice a day. In the harvest season, the Jino people throw the first handful of new rice into the fireplace to pray for another prosperous harvest year. The forms and contents of fireplace worship vary from nationality to nationality, its notion however remains the same: that the happiness, peace, fortune and mishaps of people are at the mercy of the god of the fireplace. The harvest of crops, health of animals and people are credited to the good blessings of the god of the fireplace, whereas adverse situations, particularly houses burning, are seen as the fireplace god's punishment for an offence. It is critical for the Jino, Bulang and Hani people to reset the fireplace to worship the god of the fireplace after a house burns down. The Mosuo people even hang a bundle of hog fat as a weapon to crack down on the devil god of the fireplace. Obviously, there are good or bad gods of the fireplace, and most nationalities pray for the kindness of the god of the fireplace. A unique practice of the Mosuo people is the beating of the god of the fireplace with some filthy substance.

It is well known that among many nationalities it is fire which brings in happiness and light to humankind, and functions as energy to warm and cook food as well as to slash-and-burn fields. Fire also brings disasters, the worst of which is the burning down of house and property. All this augments people's consciousness and respect for fire and the fireplace. The Dulong people believe that the fireplace is a direct connection to heaven from where the god of the fireplace comes.

The fireplace is closely associated with societal events and cultural activities. In the occasion of building a new house and during the national spring festival, the Wa and Lahu people surround the fireplace singing and dancing. It is also the venue for young women and men of these nationalities to exchange their feelings, leading to marriage. Young people of many nationalities, after their coming-of-age rituals, are allowed to date each other at fireplace celebrations.

Gender Hierarchies at the Fireplace

The status and gender hierarchies of different nationalities are also reflected at the fireplace in the arrangement of the seats and beds beside the fire. The fireplace is the center of people's daily life and reflects patriarchal, matriarchal, and mixed societies and cultures, as well as personal relationships. Visiting the families of various nationalities, one can see the seats beside the fireplace as a map of gender relations, seniority, host/guest relations and so forth. For the Aini people (a branch of the Hani), for instance, women and men are seated separately on each side of an imaginary square around the fire. For the Dai people in Dehong, the principal man sits along the upper end of the fire and the woman dependant sits on his right side, leaving his left side for the principal guests; whereas for Dai people in Xishuangbanna, parents of the host sit in these places at the upper end of the fireplace. For matriarchal Mosuo, however, the upper end of the fireplace, which is next to the ancestor shrine, is the exclusive seat of the senior woman. The remaining upper seats are also only for women and are arranged in order according to age, with men seated in the lower part of the fireplace. Mosuo women's ownership of high status fireplace seats clearly reflects their high status in Mosuo society and culture. The high status of women is also indicated in the belief that the area around the fireplace (Yimei) is the heart of the family, where the senior woman receives visitors and watches over her female ancestors.

Matriarchal families make up the majority of the households in Mosuo society. Under their marriage system, the man normally does not get married to form a stable family in his home, but practices Azhu or "visiting marriage" instead. Men live with their mothers, and when dusk falls, they visit their woman lovers in the lovers' residences and return to their mothers' home before dawn the next day. Women entertain their male lovers in their own residences. This visiting marriage system is a lifelong practice for most Mosuo people. Women nurture their children after giving birth, and they live in their mother's residence. The father does not have the obligation of nurturing his children, but does receive a nominal designation of the father. In such a household, women have some real power. Women have their own residence to entertain men whereas men have no place to sleep but their mothers' homes. The senior woman in the household who manages family life is especially revered, and therefore is seen to deserve a superior location at the fireplace.

By contrast, among most other nationalities, the high status seats at the fireplace are possessed by men. This is true of nationalities such as the Wa, Yao, Yi, Bai and Dai in Dehong, Jingpo and Tibetans.

This arrangement is clearly associated with the patriarchal nature of these societies. In the Tibetan families in Aba, for example, the location opposite the mouth of the fireplace, into which fuelwood is fed, is the primary seat and reserved exclusively for the men hosts. In the past, the wives of chieftains were not allowed to sit in upper seats even in their own residences, but were relegated to the lower corner to the left of the men hosts.

In Naxi families in Yangyuan county in Sichuan province, there are two fireplaces—upper and lower. The ancestor shrine sits behind the upper fireplace flanked by the "Principal's Bed" for the family patriarch. The bed to the side of the Principal's Bed—the "Dependent's Bed" is for the senior woman and also for kids and young men to sit and enjoy their tea. The lower fireplace is never lit in summer and is only used in winter for heat supply, boiling water and cooking by the daughter-in-law and unmarried children. The meal for the whole family is cooked at the upper fireplace. Occupying the holy seat by the upper fireplace allows the family patriarch to exclusively own the ancestor oracles and the symbols of the family. While the daughter-in-law, adult women and children have the lower fireplace for cooking, heat supply, and eating meals away from men, the Dependent's Bed at the upper fire can be shared by men. This phenomenon reflects the gender hierarchy in Naxi society, and the fact that the Naxi communities in this region are patriarchal societies. The father takes care of family affairs, and the oldest son inherits this status after his father's death. All the significant events in the family such as marriage, death, house building, ancestor and heaven worship, the household's annual production plan, its income and expenditure are in his control.

In contrast to the Naxi and other patriarchal nationalities, marriage rules in Dai villages are matrilocal; they require that a man lives with the wife's family. This means he cannot take charge of the household without establishing a separate household after marriage. Normally, the wife dominates the family, and the husband is merely an outsider. The wife controls the household economy, finances and property. Women also prevail in market activities, and 90 percent of Dai salespersons are women. Women's economic dominance in Dai households inhibits the authority of men. Men however do hold privileges in the ownership of political and religious rights. Thus, men and women have their relative authority in society and culture, which is reflected in a more symmetric arrangement of seating and fireplace location.

The Fireplace and the Founding of Households

The intimate connection between the founding of households and the fireplace is widely prevalent among nationalities in southwestern China. When the primary fireplace is set up and the lighting ritual is held in celebration of a newly built house or a small family unit derivative of their parent family, the significance of the fireplace goes beyond being merely a tool, it in fact becomes a powerful symbol of the family unit.

For the Dulong people, in fact, a newlywed couple, instead of building a new house, sets up a new fireplace in their parent's residence to indicate their small family unit. Some families have several fireplaces, meaning parents in these households live with several married children. In such large families, although the members still work together and share their food, their lifestyle is somewhat different from the families with unmarried children since they have a number of small family units each with their own fireplace. The Dulong people allocate and distribute harvest yields based on the number of fireplaces in a household. A given family unit may work only with certain other families in cultivation, but the yield of their labor is shared among all individual families with a fireplace. Cooking is similarly rotated through each of the respective family fireplaces, those not used remaining idle until their turn in the rotation. Food grain is also believed to be given birth to by each of the fireplaces in turn. A marriage relationship is not thought to be officially established until a couple sets up their own fireplace, which also indicates the establishment of an independent, small household, in terms of economy and consumption.

Large families of the Jino people in Mangshi village follow a pattern which is contrary to this form. They have a single fireplace in their household, symbolizing one unified family. A newlywed couple is not entitled to set up their own fireplace even after they have married. While people work together and share the yields, young couples are not economically independent as long as their parents are still alive, and children are not allowed to set up extra fireplaces. Even when some of the Jino families in the 1940s totaled over 40 members, the children would not set up separate fireplaces or move out of the family household until after the death of their parents. In this paradigm, the establishment of a new fireplace means the birth of a new family, which subsequently translates into the disintegration of the old big family. This explains why it is forbidden to set up a separate fireplace while the parent is still alive.

In Yanuo Jino village in the 1950s, however, where there were only a few big family households, each accommodating up to 25 smaller families. There was a general fireplace for common use and a large number of fireplaces representing each small family. The small family with its own fireplace was independent in production and consumption, and was entitled to use clan land for cultivation. There used to be limited intermarriage within the clan, but some people made unofficial marriages in the big house by cohabiting all the time. In this case, however, they were not allowed to set up their own fireplaces, and had to share their parents' fireplace or inherit it after their parents were dead. Likewise, they were not allowed to share land within the clan community nor participate in religious activities or democratic decision-making without fireplaces of their own. The absence of the fireplace meant small quasi-families were not recognized in the sense of independence and traditional rights.

The Fireplace and Clan Lineage

A new family is derived from the preceding family by building a separate fireplace, which eventually formulates a clan. Although people live an independent life with separate fireplaces, all derivative fireplaces reflect the life of the lineage of the clan.

Hani villages in southern Yunnan province are generally composed of a couple of clans, and each clan has a specific family tree. People are accustomed to remembering where they come from in terms of which fireplace their ancestors belonged to. A general fireplace is set up in the residence of each senior member of the clan to represent the clan lineage. Religious rituals are held during annual festivals to worship the general fireplace of the clan, praying for peace for the entire clan from the clan deity.

The symbolism of the fireplace in relation to the clan can be further tracked down through the evolutionary process among the big patriarchal clan families of the Deang people. In their big families, even if the son has a wife, he is not supposed to marry while his parents are still alive. The big thatch house is divided into a number of smaller partitions which are called *Geduo*, one for each couple. Other unmarried children live in their parents' section of the house. The big house has two fireplaces, one for cooking and the other for heating water. A couple living in their partitioned section has no separate fireplace. Everyone works and eats together around the main fire. After the death of the parents, sons with wives can then make separate residences out of the house and set up independent fireplaces. When a new fireplace is built, the fire must be acquired from the dead parent's fireplace, meaning they are derived from their parent through the fireplace. Throughout history, the Deang descendants have derived and preserved their clan lineage from their patriarchal ancestor. Although the fireplace of the clans' first ancestors no longer exists, the fireplace of the senior of the clan is assumed to be the general fireplace of the clan, and all clan members are supposed to cook fresh rice in the general fireplace and worship it during the first tasting of new rice.

Entering the Tibetan wooden houses on the border of Yunnan and Tibet, we can see a fireplace on the ground composed of three polished pot-posting stones. These three stones are regarded as the "holy things" of the family. Once they are buried, they cannot be removed again. The pot-posting stones in many families today have been used for several generations. When a new family comes into existence, the fire comes from the fireplace which their ancestors used. When the fire goes out in a family's fireplace, the people will have to regenerate it from the fireplace of their ancestors. It can be seen from this habitual rule that when people generate fire from their ancestors' fireplace, they proclaim themselves to belong to the same patriarchal ancestor, in addition to the family tree which further distinguishes who they are and where they come from. Like the Tibetans, when the Dai people living in Ruili of western Yunnan build a new house, the senior woman lights the fire for the fireplace while lauding the origin of the clan to commemorate their ancestors. In this ritual, the fireplace is a also a symbol of the clan ancestors.

In Yanuo village of the Jino, people live in big longhouses by clan. With an increase in population, when longhouses built on mountain sides cannot be expanded to accommodate more people, the clan has to separate to live in several big longhouses. The Azerao clan, for example, had four longhouses in the past, including the Mamoaima longhouse, which accommodated up to 125 people in the 1940s. With time, however, people are aware only of their ancestors' general origins rather than specifically defined kin. People worship the fireplace of their ancestor and attribute to themselves kindred roots. In big longhouses, the first fireplace next to the entry represents the general fireplace of the clan with the small fireplaces located away from it in evolutionary sequence. The general fireplace is owned by the entire clan rather than any small family. In the Azerao clan case, for example, the general fireplace was owned by the entire clan despite the existence of four longhouses. Customs pertinent to the general fireplace reflect kindred and clan relationships.

The Fireplace, Food, and Sustenance

In the worship of the deity of the fireplace, people believe the deity can ensure basic sustenance and the continued existence of a people, and there are thus many important religious rituals to propitiate the fireplace deity. Another obvious reason why people believe in the importance of the fireplace is that this is where they cook their daily food, explicitly reflecting the deity's function in protecting the fireplace as food, the most essential thing in life.

The Mosuo people by Lugu lake regard fire as the symbol of light and prosperity, and the fireplace as the sign of sustenance. The flames of a fire are therefore vigorously stoked with wood each day. It is said that the fire in the fireplace of some families has been maintained for generations. A nicely burning fire with bright flames and sparks flying from the fireplace is considered to be auspicious, symbolizing good luck in the future, plentiful harvests, and healthy animals. Conversely, if the fire is feeble and full of smoke, it forecasts bad luck.

The Yi people in Yjao Liangshang fry buckwheat flowers as a ritual to pray for a good harvest. After a rock is heated on the fireplace, buckwheat seeds are thrown onto the hot rock to make them expand. The more extensive the explosion, the bigger are the flowers and fruits of harvest.

With the arrival of the harvest season, the Jino people select an auspicious day to read the fortune of yields for the next year by the fireplace. At this time, the whole family sits around the fireplace and lights boughs of wood, observing where the smoke emerges from the flames. If the smoke emerges from the left side of the fire first, people believe that they will have a good harvest in the coming year. If it comes from the right side, this means they will be able to hunt sufficient animals in the next year. However, if smoke emerges from behind the fireplace, this indicates impending sickness in the family. After this ritual, the people of the households start reaping crops in the fields.

The Jino people in Yanuo village, on the first day of the fall harvest, collect new grain from the field of the clan senior and cook a fresh rice meal on the general fireplace of the clan. Each small household contributes its rice to be collectively cooked on the general fireplace. By this process, people intensify the fire and flames which light the corridors of the big house. They hope that life stays as warm as the heat of the fire, and do not care if the rice is scorched in the process. After the cooking is done, the senior clan member grabs a handful of fresh rice and throws it into the general fireplace murmuring appreciation and asking for the blessing of the god of the fireplace for abundant grain. He then throws a handful of fresh rice into the fireplaces of each small family praying:

> Those who have new rice in their fireplaces are of the same family, may the God of the general fireplace bless each fireplace with flames as high as the general fireplace everyday and bring prosperity to human beings and animals, good harvests and health for all living beings.

After this ritual, each small family is entitled to cook new rice on their fireplace and happily taste what they have earned throughout the year.

The Fireplace and the Rituals of Life

Many nationalities such as the Jingbo, Yi, Jino, and Mosuo, worship the fireplace during various festivals to pray for fertility and a prosperous population. The Lisu people in Yongsheng county believe that the deity of the fireplace controls women's fertility and the birth of children. They create a strong fire in the morning, as in their minds, the stronger the first fire, the more prosperous their children would be in the future. When adding wood to the fire, the head of the wood should go in first, followed by the tail to ensure the successful delivery of a baby. The reverse sequence would mean difficulty in giving birth i.e., the baby's feet would come out of the womb first instead of the head. The firewood used should also be smooth and regularly shaped to help in a successful delivery.

The Yi people in Dongchuan and Xuanwei believe that when a pregnant woman encounters difficulty in giving birth, the *Bimo* (shaman) should be invited to recite the holy texts of the god of the fireplace to pray for a successful delivery. After the ritual, the *Bimo* puts several smooth pebbles into the fireplace to heat them and then places them into a water basin floating with pieces of cypress. He then picks up the cypress and spins it thrice over the head of the woman giving birth. This assures that the deity of the fireplace has attached itself to the woman's body. Subsequently, he takes a piece of charcoal out of the fireplace and soaks it in a bowl of water for the woman to drink. After this, it is said that the woman will give birth successfully. This ritual shows the firm association of the fireplace with birthing.

In Yanuo village of the Jino, if any household in the big house has a birth, a month after the birth, the father will roast a hen to ask for the blessing of his deity for his children. After the roasting is finished, the hen is circled over the fireplace several times, and the man eats it on behalf of his ancestors. During this ritual no one is allowed to talk to him nor can he talk to anyone, otherwise the ancestor deity will not enter his body.

The coming-of-age ritual of the Purni people also revolves around the fireplace. On the first day of the Purni's national holiday, youngsters over 13-years-old participate in this ritual. Armed with earrings, linen cloth and ring balls, a girl has to jump and land with her feet on two bags: a bag of hog fat and another of grain on the right side of the fireplace. Hog fat is the sign of fortune, and grain is a sign of harvest. The mother acts as the master of ceremonies. The shaman in turn prays to the fireplace and the family ancestor shrine, and then the mother takes off the girl's linen gown, dresses her in a linen coat and crepaline shirt, and ties a patterned belt around her waist. The newly dressed girl has to bow before the fireplace and shrine of the ancestor. If the coming-of-age ceremony is for a boy, he has to stand to the left of the fireplace, similarly jumping and landing with his feet on bags of hog fat and grain. He is armed with a sword and silver, and salutes, bows and prays to the fireplace and the shrine of the ancestor. People believe that by bowing to the fireplace, one can be blessed with food and clothes, flocks of animals and maximal bins of grain, while bowing to the ancestor shrine can be a blessing for good harvest of grain and crops, and healthy children.

Another example of a coming-of-age ritual is the "wearing trousers" ceremony of Yi boys in Xiaoliangshan, which reflects a typical relationship between the fireplace and the family. The mother heats a piece of stone in the fireplace and pours a gourd ladle of cold water over the stone. When the vapor emitted by the heated stone forms a circle, she puts a pair of new trousers on the boy. The boy is now considered an adult with blessing acquired from the god of fireplace. Whether he marries or not, the blessing of the family spirit is then always with him.

In Yanuo village, Jino family relationships can also be seen in the role of the fireplace when a girl from a big longhouse prepares for marriage. On the morning of the wedding day, the groom sends gifts to the entire family of the bride. The groom and his friends wrap a number of gift bags filled with meat, tobacco leaves and dishes wrapped in banana leaves and then place them beside the tripods of the fireplaces of the bride's entire family. At the evening wedding banquet, a big pot of rice meal is cooked on each fireplace of the bride's house to provide opulent dishes for the guests. When the guests come, the senior host and other dignitaries sit beside the general fireplace, and the other guests surround each respective fireplace to drink, eat, and have fun.

For weddings among the Yi families living by the bank of Jingshajiang (the upper tributary of the Mekong river), the side of the fireplace against the wall and the pot-posting stone on this side represents the ancestors. The left pot-posting stone represents the young man and the right one the young woman, set in pairs to symbolize continuity of descendants in the future.

As a unit a Yi family also prays for good harvest and peace centered around the fireplace and its worship. So long as a family is still in existence, the fireplace cannot be removed. The reincarnation ritual of a bride of the Yi people in Yiao Liangshang offers us new evidence for the close association of the fireplace with the family. On the second day of marriage, a religious ritual for the reincarnation of the bride's soul is held at the groom's fireplace presided over by the *Bimo*. The bride and groom including the groom's family sit separately in seats on either side of the fireplace. The *Bimo* lifts a bound live female goat and spins it seven times around the head of the bride from the left, and then nine times around the head of groom from the right. She then kills the goat and drips the blood into the fireplace. In her prayer for the bride, the *Bimo* recites: "Your soul has been reincarnated from your parents' fireplace to this fireplace. You will have your meals with this fireplace. May you grow strong and healthy with these meals and quickly have children." The bride then officially becomes a member of her husband's family after this ritual, which ends with the transformation of the bride's soul to her husband's family at the fireplace.

The Culture of the Single Fireplace

Among most nationalities using the fireplace, it is common for one family to use only one fireplace. We can group into this category nationalities such as the Miao, Yao, most Yi, Dai, some Bai, Naxi, Hani, Bulang, and Deang. Single fireplace households are patriarchal units of one couple or small families. In such families, all the members work and have meals together, and when their children get married, they make separate households for themselves. For the Hani people in western Yunnan, a married son has to separate himself from his parents to make an independent fireplace and household, even if the separation is still inside the big courtyard. After establishing a separate household, his new family is relatively independent economically, but cultivated fields are not necessarily separated, and they will share grain with their parents after harvest. All sons make separate households after marriage except the youngest son and his wife who stay to take care of his parents. In this case, there is only one fireplace in the household.

There are big families with single fireplaces among the Hani, Deang and Yao people in Honghe. The origin of these big families has nothing to do with the residence, but they have been formed because of the traditional rules governing prohibition of division of families while the parents are still alive. For instance, in the big family of the single fireplace Hani, three or four generations usually live together and couples are dependent upon their income and consumption on the rest of the family. The production of the family is arranged uniformly by the father and the senior son, while family life is in the charge of the mother and the senior daughter-in-law. In such a big family, a son making an extra fireplace is deemed as a betrayal. Only after the death of the parents can the division of the family be conducted by the senior son and his wife. In this, the entire family gets a share of the possession, thus small families are allowed to set up their own houses with an independent fireplace life.

Gender Divisions and the Dual Fireplace

The dual fireplace exists in families of many nationalities with differing cultural notions and functions relevant to the specific cultural background of each nationality. For the Dai in Mangshi, in addition to a fireplace for cooking, they also have a fireplace for heat supply in the sitting room. The kitchen is separated from the sitting room and living room and used exclusively for cooking purposes, while the sitting room is the center of daily life.

The dual fireplaces in the big bamboo houses in Zhengkang of the Deang people are also associated with family structure. The family size of the Deang people in this region in the past ranged from over 10 to 50 members. These big families derived from the descendants of a male ancestor. Children cannot move out of the big house if their parents are still alive. Marriage and birth occur in the same house. The consequence of this convention is that the children still live in the same building to show intimacy even after their parents are dead. The Deang people value this family culture in life for harmony and mutual assistance. One particular big long-house occupied 500–700 m² in the middle of the 20th century. In such a large family, with multiple couples and their children, people were economically dependent, and worked and ate together. But in the past there were no separate fireplaces for small families whatsoever, and the whole family used a single shared fireplace. However, since a single fireplace was barely satisfactory to meet the needs of many people, the dual fireplace eventually came into being. One fireplace was then used for cooking rice and another for cooking main dishes. In this case, the emergence of a third fireplace, however, indicated the disintegration of the big family.

The dual fireplace exists in families of the Jino, Jingo and Wa people with distinct roots in the need to meet growing household populations. In this case, one fireplace is used for cooking and heat supply, and the other for cooking hog food. Another dual fireplace is associated with the aforementioned gender issues. A typical example is the big bamboo building with railings among the Aini people in Xishuangbanna. The interior of the big building is partitioned into two interconnected big rooms. Each room has a separate fireplace and staircase. The arrangement and layout of fireplaces is directly related to the gendered cultural practices of the Aini people. The two partitioned rooms are for women and men respectively. The men's room is called *Baolou* and the women's *Longma*. Married women and men live in separate household spaces in a family before a couple is entitled to set up their own independent residence. The father lives with his associated gender regardless of the marital status as does the mother. Gender-mingled living is forbidden, even for aged couples as a rule. The layout of this case is different from those of other nationalities with dual fireplaces. Here the gendered fireplace in each room has distinct functional and cultural notions. The fireplace in the women's room is for cooking and boiling water while that of the men's room is regarded as the residence of the god of the fireplace to be worshipped very often. The latter also serves as a guest reception area in addition to being used for daily heat supply and for boiling water for men.

In the Wa families in Haidong of Menglian county, fireplaces are similarly categorized into a women's fireplace and a men's fireplace. In the culture coexisting with this dual fireplace, both genders have substantial power. Although the Wa nationality is assumed to be a patriarchal society, with men dominating activities of production, war and religion, women are superior to men in power in many dimensions. For instance, the big house of a family is named after the woman proprietor, and the ownership of family possessions is taken over by the in-house daughter taking care of the parents. In the harvest season, the senior woman directs the serving of new rice. If the senior woman has passed away, the respected women of the village are invited to do this. Women in this region also control the religious activity of fortune-telling with hens.

Obviously, the rights and obligations of Wa women and men are different from other nationalities in this region. The strong proof of this paradigm is that there is a unique convention in the Wa families of this region of allocating children to their mother or father. After the birth of a child, the wife and husband have to define the allocation. In general, the wife receives more children than the husband. Traditionally, children allocated to the mother have to live with their mother's family before and after her death, they have to be buried in the grave of their mother's clan. The reverse applies to the children belonging to the father. This definite classification of the ownership of the children in terms of the immediate family, clan and grave to which the children belong is in fact the projection of gender groupings in societal practices. The dual-fireplace characterizing the genders is associated with this societal culture, as also the origin of the dual fireplace among the Wa people in this region.

There is another gendered dual fireplace system in the villages of the Deang people in the Sangdai mountain of Luxi. Two fireplaces are arranged in their bamboo houses. One is the internal fireplace *Genawa* and the other is the external fireplace *Le*. The interior fireplace represents women and is exclusively for them, whereas the exterior one represents men and is exclusively for the father and unmarried men. Women get their heat supply in the daytime and sleep at night from the internal fireplace. Men get their heat supply from the external fireplace. They receive visitors and sleep next to the external fireplace.

Another dual fireplace of the Wa nationality with different cultural notions from those in Menglian is in Ximeng. Here, the bamboo houses of the Wa are generally split into two rooms with two doors each, opening towards the outside. The two rooms are interconnected with a door; one room is for the host and the other for the guests, each room with one fireplace. The host room is for daily rest and sleeping for family members, and the associated fireplace is for cooking and heat supply. The guest room is exclusively for entertaining guests with the associated fireplace only for cooking and heat supply for the guests. Essentially, it is also a spirit fireplace for exclusive worship of spirits by family members. In the event of birth, sickness, death or evil omens, people have to worship the spirits in this room. Conceptually, people believe that spirits and souls exist in everything i.e., there is a heaven spirit in heaven, an earth spirit on earth, a water spirit in water; everything is under the control of spirits. Hence, the Wa people worship spirits for everything and ask for their blessings to fulfil various needs. However, there are good spirits, and bad spirits, and this is why people have to separate the fireplace for daily cooking from that for spirit worship. The host fireplace for daily cooking symbolizes sustenance and working effort, and blesses people with food. It is thus classified as the "good" fireplace whereas the spirit fireplace controls sickness, pain, birth and aging, and is therefore associated with "evil" and has to be kept separate from the good fireplace.

The dual fireplace also exists in Naxi communities in the Erya region of Muli county in Sichuan province as well. The cultural implications of this dual fireplace is reflected in status hierarchies. In this case, the sitting room is the venue for cooking, decision-making, and guest entertaining with its fireplace perceived as the center of a family. In the sitting room, there are two fireplaces conceptually arranged on upper and lower sides. The upper fireplace is arranged on a platform 1 m in height with two beds on both sides, complemented by one couch each. Though there are two fireplaces, the true symbolic fireplace for the whole family is the upper one as the divine stone of the ancestors and the pot-posting stones of the god of the fireplace kept are here. The fire in the upper fireplace can never be allowed to go out, and all worship activities are conducted at the upper fireplace. The principal women and men sleep on either side of the upper fireplace, respectively, with 11 "authentic beds" for the men and "auxiliary beds" for the women. The two sides of the lower fireplace have corresponding couches indiscriminately arranged for both women and men. There is one exception: no woman but the principal woman can sit on the couch beside the upper fireplace. The other women have to sit on the couches beside the lower fireplace. Also, neither the children nor adult women are allowed to sit on the couches beside the upper fireplace, implying a certain degree of gender discrimination.

III

Shu: Naxi Nature Goddess Archetype

XI YUHUA

There are currently 280,000 Naxi people in southwestern China, 68 percent of whom are concentrated in Lijiang Naxi autonomous county in Yunnan province. There are also Naxi in Zhongdian and Weixi counties in Yunnan province, and Yanyuan and Muli counties in Sichuan province. Before the 1950s, the Dongba religion was the principal faith of Naxi regions. While Dongba is primarily based on ancestor worship, it is infused with the spirit of Animism: all things in nature are believed to have feelings and may suffer from hunger or become satiated, feel cold or hot and are capable of bad or good behavior. In the animist perception of the universe, *things* are the receptacles of *spirit*, which animates trees, rivers, and mountains. *Spirit* also determines the shift of the seasons, night following day, the natural rhythm of life that leads to aging and death, and the alternation of wind and clouds.

From the generalized perception of "spirit" comes the more specific concept of nature "spirits" or gods, called "Shu" by the Naxi. Shu are subject to worship while being distinguished from other categories of spirits with respect to the manner in which they are worshipped. Contact with Shu is subject to the same level of equality as contact between different people, and the give and take between "worshipper" and "worshipped" thus becomes similar to the give and take of rights and privileges of members of the same household. The community *takes* from Shu as part of its daily life—cutting down trees, hunting and storing and using water, and *gives* in the form of religious ritual as a means of maintaining the natural balance between it and its environment, and therefore between its members and Shu. For this reason, the most important of all rituals of worship, which takes place at the village level in the second month of the lunar calendar, is called "repayment of debt to Shu." Both this ritual and its name are a metaphor for the most basic principle of Naxi nature worship, that humankind must control its appetites, not taking more than it should; the observance of such a principle means that the more a community takes from its environment the more it must return to its Shu.

In this chapter, we use the concept of "archetype" developed in Jungian psychology. Jung believes that human cultures contain certain basic, primordial conceptual types he called archetypes. These types do not represent specific forms existing in chronological and/or geographical space, but are both direct projections of the human psychological landscape and images instrumental to its construction: they invariably manifest themselves in mythological form, leading to the creation of gods and spirits. For this reason, spirits in many of the world's religions tend to be infused with symbolism, a repository of symbols that reflects the progress of collective human psychology. Our own analysis of mother goddess archetypes in Naxi religion is chiefly derived from views espoused by Erich Neumann in *The Mother Goddess—An Analysis of the Archetype* (1963).

As Neumann himself says: "The prototype in its unforgettable symbolic imagery has a wealth of meaning and is full of life and mystery in a way reminiscent of the primary importance of instinct to human existence" (ibid.: 5). Neumann quotes psychologist Jung as saying:

> Primary imagery can be appropriately described as instinctual self-awareness or the ego's portrait of human instinctual processes at work.... Archetypes are in essence 'indescribable,' representing an 'inclination' or 'inclinations' expressed in a particular form using material from various levels of the human consciousness and performing a role at a particular phase of the development of the human psyche (ibid.: 6).

According to Neumann (1963), the prototypical source of mythical archetypes, springing as it does from the subconscious mind, is by nature chaotic, obscure, and infused with contradictory, even diametrically opposed content. Thus, the ying-yang dualism inherent in early cosmology is, in its prototypical form, not a neatly divided graph but a muddle of contradicting elements—feminine and masculine, good and bad, etc., all blended into a single serpent-tailed primary image, an image Neumann calls "Uroboros." The perfect manifestation of the Uroboros is characterized by contradiction and conflict, uniting unrelated elements by fusing or attaching them together and thus forming a single, prototypical image.

"We have already mentioned that within this initial phase of human psychology, masculine and feminine, the conscious and unconscious, and various other Uroboros elements mix together" (ibid.: 216). Within the transitional stage from unconscious to conscious, the feminine elements of the Uroboros take precedence in spite of the lack of a strong sense of self-awareness. At this point, numerous, undifferentiated and ambiguous duality evolves into a new phase of development, characterized by the emergence of female archetypes.

After an examination of Neumann's theories concerning Uroboros and the emergence of feminine archetypes, one is naturally lead to draw conclusions concerning the Shu of the Dongba religion. Shu, like the Uroboros, are characterized by intertwined serpentine tails; in its fullest manifestation, the Shu type of nature spirit has the head of a frog, the body of a human, and intertwining serpentine tails. The most important feature of this symbolic, nonhuman representation of Shu (because of its place at the top of the total image, its "head") is its feminine frog head, ultimately attached to masculine intertwining serpent tails. Nor is the occurrence of a feminine archetype an isolated one; religious belief, historical and cultural background, an examination into "Shu," with reference to the concept of archetype as Neumann defines it, allows us to explore a very representative example of the symbolic significance of feminine archetypes.

Shu Symbolism

Other Naxi terms for the same nature deities most commonly referred to as Shu include Limu, Shi-er and Chadao. "Shu" refers to the category as a whole, while the other terms refer to specific subtypes or aspects of the more generalized "Shu." Shu means life or life force, liveliness or abundance (prospering life). Limu means the mother of the earth, Shi-er is Naxi for a snake after its skin has peeled, and Chadao is a Tibetan loanword referring to an earth deity. Taken as a whole, these terms represent a range of interconnecting images including earth or land, mother, snake and deity.

Dongba sacred texts provide us with a gender-based explanation for the peculiar three-in-one image of the Shu—frog head, human body, and serpent tails. One such text says: "The snake-headed ghost of badness is the father of the ghost of smelliness; the long frog-headed smelly ghost is the mother of the smelly ghost" (He Shicheng and Li Jingsheng, 1981a: 9). The "smelly ghost" refers to a spirit of badness or filth. According to another text: "The long armed woman did not stab the snake, the long armed man did not tread on the frog" (He Shicheng and Li Jingsheng, 1981b). The reference here is to people not doing things that harm nature, such as killing snakes and frogs, and thus not offending the nature spirit or spirits. The first and second line approach the problem of man's opposition to, and unity with, nature from different yet fundamentally related and interacting perspectives: the image of women stabbing snakes is used because snakes represent their gender opposite, just as the image of men treading on frogs is used to oppose men to their symbolic feminine opposite.

More evidence of this symbolic reading of gender opposites can be found in a story from sacred texts involved in the worship of Shu. According to the story, once while Duxia-ao was out sheepherding deep in the mountains, he discovered that Niushaxuluo was having an affair with his wife Duzhouxuma, and rushed back home. When Niushaxuluo saw that Duxia-ao was about to enter the house, he changed into a snake and hid in a box (He Yuncai and Li Jingsheng, 1981). Duxia-ao, son of a Shu, also takes the form of a snake.

While the evidence from Dongba texts is enough to support the view that Naxi religion views snakes as a symbol of masculinity and frogs as a symbol of femininity, additional evidence from the structure of the Dongba pictographic script supports our reading of Shu as a feminine archetype. As Fang Guoyu and He Zhiwu (1981) note, the script is distinguished by a tendency to emphasize the "key element" of a given graph:

> In case of graphs whose basic structure and appearance are similar, the key element of a given graph is emphasized, so that while different graphs of the same type are very similar, the

difference (in emphasis) is sufficient to keep the reader from getting them mixed up.

In practical terms, this means that the graphic element put on the top or "head" of the graph tells us the "identity" of the graph as a whole. As a tiger head represents a tiger, a chicken head a chicken, and so on, it follows that the frog head of the Shu is a prominent feature and means "frog," thus linking Shu with femininity as part of its primary aspect or identity, while the masculine serpentine tails, being located on the "tail end" of the Shu merely represent a secondary associated feature. The graph as a whole, links ascendant feminine identity with secondary masculinity by means of a human body, thus giving us a complete image of a Uroboros—related prototypical feminine archetype. This puts us in mind of what Neumann has to say about early feminine archetypes:

> The snake is an associated phallic element and either becomes a part of the feminine archetype or her mate or companion. Thus, in (Minoan) Crete and in India, the snake is an aspect of the feminine deity represented and, simultaneously, her phallic-masculine companion (Neumann, 1963: 145).

Texts used in the worship of Shu explain why Shu are worshipped:

> Haizibian, on the vast earth, went before Shu Chouluchou (*lu* is derived from a Han Chinese loan word for dragon, "Long") and said: "If *fuzhe* is not enough, one prays for more *fuzhe*, if one is short of offspring one prays for offspring, if there is not enough water from the skies one prays for rain, if the crops in the fields are insufficient one prays for abundant crops, if one's livestock in one's stables are few in number, one prays for livestock, if there is a deficiency of livestock deities one prays for more livestock deities and if there is too little grain in the barn, one prays for more grain" (He Shicheng and Li Jingsheng, 1981a).

Of the items being prayed for, *fuzhe* represents the collective good fortune and peace of a population, offspring are necessary for maintenance of the human race, and water assures human livelihood, making the last of these directly responsible for crops, livestock and grain. In a word, praying to the Shu assures the fertility of all things. Shu live in the water while governing both the waters in the sky and upon the surface of the earth, while water nurtures all life. Rainwater is associated with femininity: According to myth it comes from an egg laid by a great yellow frog living in the water.

Water in the Dongba religion is femininity impersonated and has a tremendous life giving force:

> Meilidongzu lived on the white earth. One day, Meilidongzu went to the sea for a walk and stood there, seeing his lonely shadow in the water. He then wished for a companion to be forever at his side and spit white spittle from his mouth into the water saying 'Let the true, the good and the beautiful float to the top while the false, the bad and the ugly sink to the bottom.' In three days, a green agate-like woman of great beauty appeared in the water (He Yuncai and He Pingzheng, 1981).

Another version of this story goes as follows:

> Meilishu lived on the dark earth. One day, he went to the seaside and stood there, seeing his lonely image in the water. He then wished for a companion to be forever at his side and shed tears into the water. As he did so, he uttered the words: 'Let the good and beautiful rise to the top while the wicked and bad fall to the bottom.' After three days, a beautiful woman who was like black agate rose to the surface (ibid.).

The principal goal of Shu worship is to ask for rainwater and off-spring. Offspring must come from the primary matrix or "mother body." Water is the source from which all life is nurtured. As the Shu, which control the rain and are full of life-giving power, are closely associated with femininity, the object of Shu worship is obviously a feminine deity, making Shu a feminine archetype.

> Give medicine to the 99 white Niugedunwu Shu that live in the skies. Give medicine to the 77 Nuishashuluo Shu that live on the earth. Give medicine to the 55 Shu in the mountains, give medicine to the 33 Shu in the tree-filled mountain valleys. Give medicine to the Shu on the banks of the river, in the nine trees and nine stones and along the banks of the nine small streams. Give medicine to the Dungesibei on the 12 mountains, the 12 plains, the 12 cliffs (A Complete Annotated Translation of Naxi Dongba Texts, 1981: 175).

Note that the passage begins with the numbers 99 and 77. The former is derived from the masculine number nine, while the latter is derived from the feminine number seven. Nines and sevens go together to form the totality of Shu, thus demonstrating that Shu is a concept that includes both masculinity and femininity.

Abstinence from taking life is required during the period of Shu worship. A cock already blessed ("carrying wishes") is presented to the Shu with strings of various colors attached to its legs. After the ceremony, the cock is released into the wild as a companion to the Shu. A clue to the significance of this act is provided by the Naxi funerary custom of making offerings of animals of opposite gender to the deceased. If the deceased is a man, a hen is offered, while if the deceased is a woman, a cock is offered. By the same principle, then, it would seem that Shu are female.

The Elementary Character of Shu

Neumann believes that the elementary character of the feminine archetype can be expressed as a great orbicular container that accommodates all the things in the universe; everything springs from the Great Orbicular Container and everything revolves around it. Variations of this elementary character are biological and psychological variations of femininity. The orb itself is in fact a symbol for the womb, which is perceived as being characterized by accommodation, protection, nurturing and birth, all products of its function in pregnancy and birth. Concomitant with it is an eternal, surpassing creativity, one that, because of its being associated with this "cosmic womb," naturally becomes associated with the feminine archetype, which thus becomes resonant with womb-like, orbicular fertility. "Orbicularity" as a feminine quality thus becomes indicative of stability, permanence and great capacity.

This orbicularity is evident in the way the Shu is represented; a circular frog head and serpent tails twisting around each other in a circular fashion are connected by a round belly. Cultural change and the influence of outside cultural elements led to variations in how the Shu graph was written. If the reader will examine the pictographs in Figure 3.1, she/he will find three groupings of pictographs, the left-hand group representing "frog," the group in the middle column representing "snake" and the right hand one "Shu."

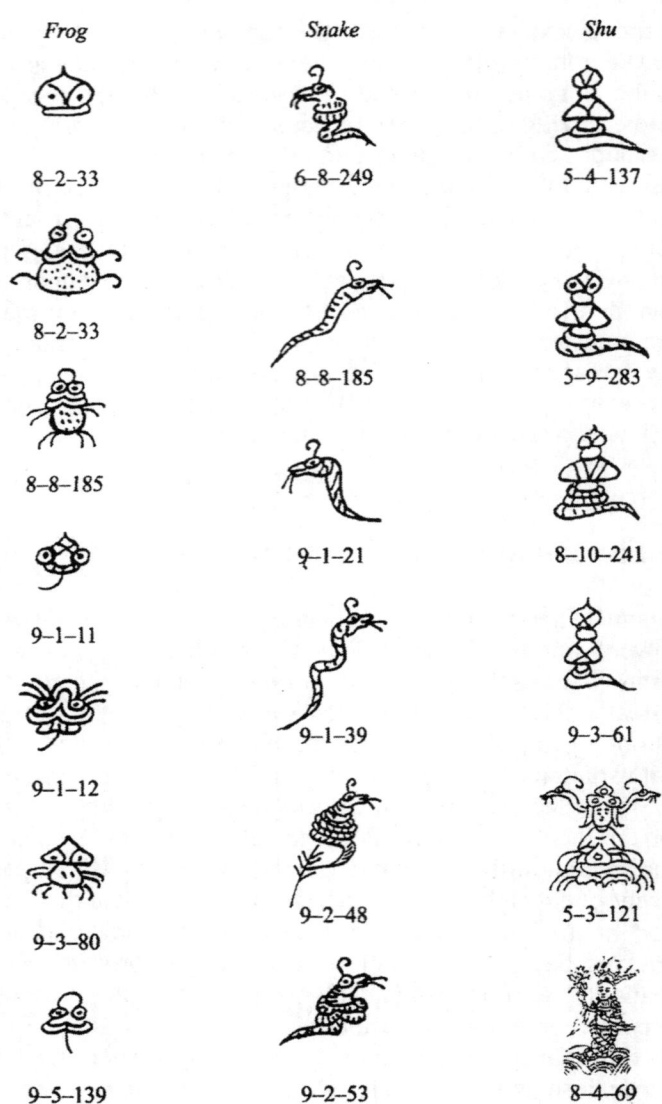

Figure 3.1
Naxi Pictographs Showing Frog, Snake and Shu

Source: *A Complete Annotated Translation of Naxi Dongba Texts* (1981).
Note: The numbers below each pictograph are a form of bibliographical reference, representing volume, book, and page number: hence "8–2–33" means "Volume 8, Book 2, page 33."

In one variation of the pictographs in Figure 3.1, a snake head appears above the head, on which is drawn a Buddha face. In another variation, the latest form of Shu, the Buddha face is rendered with sharp clarity, a lotus placed in one of the Shu's hands and a *kasaya*, the mantle of a senior Buddhist priest, draped over its shoulder, radically transforming its elemental character as a feminine archetype. Both variations of the Shu graph reflect a strong Buddhist influence on Dongba texts: neither, in fact, is commonly used. The most common variations of "Shu" are the top four in the column.

The orbicular Shu is closely associated with the origins of life (in animist terms, this includes *all* of nature and not merely animate nature). Our most direct, intuitive perception of Shu is that it rests firmly on the earth, attached closely to it by its serpent tails, as if it was fixed in the earth. This image is positively bursting with the symbolism of "possession-control-mastery." Shu "possess" the mountains, rivers, valleys, and all natural phenomena, and are the master of the great mother earth.

One special offering made to Shu is called "Liduo." Among the pictographic representations of Liduo are a snake attached to a wide-waisted, long-necked vessel, while a frog is attached to a bowl; a Shu image in a bowl, and a simplified graph of a sacred mountain in a bowl with a frog and a snake are on either side (Figure 3.2).

Liduo is performed by making a sacred mountain on a base of highland barley stir-fried noodles (the sacred mountain represented by the offering is Mount Naruoluo). The sacred mountain is roughly a meter

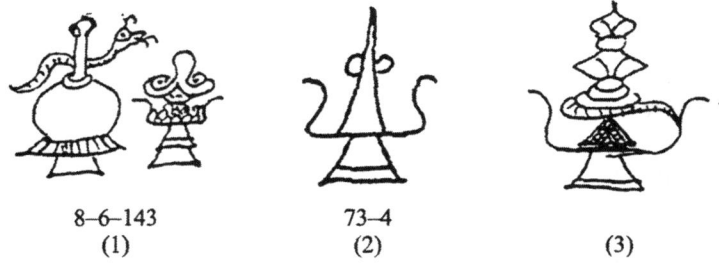

8–6–143   73–4
(1)   (2)   (3)

Figure 3.2
Pictographic Representation of Liduo

Note: Our reference for these pictographs include: (*a*) pictograph on page 143, Book 6, Volume 8 of *A Complete Annotated Translation of Naxi Dongba Texts* (1981); (*b*) Ibid. Volume 73, Book 4; and (*c*) pictograph from Fang Guoyu and He Zhiwu, 1981: p. 341.

high and 4 m in diameter, with a snake and a frog on either side of it. More than 10 similar Liduo are constructed and placed in five directions around the Shu altar. Liduo is the origin of Shu's life and the source of its form. Judging from its three-part structure—Mount Naruoluo in the center and a frog and snake on each side—it is a formal representation of Shu as a feminine archetype. Mount Naruoluo is believed to have been built by Naxi ancestors, who piled rocks into a sacred mountain after the founding of the world, slowly raising it into the sky in order to control heaven and earth. The frog and the snake are placed on the mountain, which thus becomes their residence or platform. The holy mountain rising into the sky from the earth signifies the great earth, and is thus equivalent to an earth goddess. According to another myth, when the Heavenly Father Mulao-ahpu united with the Earth Mother Cunhengahzhu they lived on the top of a mountain—Mount Naruoluo—where heaven and earth meet. The Heavenly Father's residence was above the peak, the Earth Mother's below it. When an altar is constructed in the Dongba religion, a plow is placed on it representing Mount Naruoluo. Two cypress branches or two paper flags with the sun and moon painted on them or stuck on top of the "mountain," represent the way that the bottom of Mount Naruoluo supports the earth while its peak controls the heavens.

When an altar to Shu is constructed, a wooden plaque with an image of Shu painted on it and several other wooden plaques with pictures of plants are arranged on it, after which, at night, the altar is surrounded by sackcloth, representing the dwelling place of Shu. The cloth is removed the next day when formal ceremonies begin. The painted plaques are symbolic of frogs and snakes:

> No one can carve the plaques but the Moupanenhao Dongba priest, who does so with his hands. He cannot carve the head so he imitates the head of a frog in the upstream of the Hengyiwaji river. He cannot carve the tail so he imitates the tail of a snake in the downstream of the Hengyiwaji river (He Baolin, 1981).

Another way of preparing the altar is to make a single 6 cm long snake and a single 6 cm long frog. After this is done, around 20 sets of round cakes shaped like frogs and snakes are made and placed in five directions around the altar; each cake has a diameter of roughly 3 cm.

Other offerings including eggs are used in other rituals; special offerings specific to and associated with the deity being worshipped are also made. Dongba texts contain many stories involving life hatching from eggs, even including stories about the birth of some Shu. As an object symbolic of femininity, eggs contain life and are often depicted in terms of opposing containers, such as black eggs and white eggs. We find references to eggs as an archetypical symbol of creativity in a large body of mythology. In cases where eggs are depicted as opposing or dualistic containers, they may be divided into black and white, the heavens above, and the earth below (Neumann, 1963: 41).

The Origins of Shu

A long time ago, before human beings appeared on earth, heaven and earth were connected. A change occurred in the interconnected heaven and earth bringing forth white and black clouds, white and black dew, and white and black grass. These things, too, changed, bringing forth the Meilidaji Sea, from which a single tree shoot like hair appeared. This too changed, yielding gold and silver flowers, around which blew white and black winds. The tree shoot in the Meilidaji Sea changed and produced a golden yellow egg which then produced a great golden yellow frog. The great frog lived each year in five different places on Mount Naruoluo in succession and produced, one after another, white, green, black, yellow, and multicolored eggs; and each egg produced pairs of eggs. Three years after the eggs were laid the white clouds and white winds incubated them for three days, but the eggs did not hatch. The people then asked the *Shiluo Dongba* to incubate the eggs, and from these five different colors of eggs from five different directions Shu were born (He Kaixiang and Li Jingsheng, 1981: 161–64). This myth shows a process by which Shu came from the ether, where the colors black and white produced eggs of different colors, out of which different colored Shu were hatched in different places. The matrix or "mother body" of the Shu is eggs.

The Naxi believe that the Shu emerged from eggs, an origin similar to that of all things in the universe. Eggs represent the ability of Shu to produce life, a fact symbolized by the offerings made to Shu. Such offerings to Shu as round cakes and butter are also round; these two items nurture life and feed people, functions that manifest the elemental character of Shu as a female archetype.

Sites selected for the construction of altars to Shu include gutters or pools. The altar is built on the ground, wooden plaques placed on the earth around it, Liduo placed directly on the earth as well, and many bamboo sticks, tree branches, and/or colored banners stuck on it, symbolizing the belief that Shu are bound to the earth at all times. Stone and clay are also piled together to form a 1 m tall Shu tower, around which many bamboo sticks and tree branches are stuck. On the top of the tower is a cypress branch. The tower itself is emblematic of the female archetype of the great earth.

Circular pictographs, offerings and Liduo are all manifestations of the basic characteristics of Shu, among which also exist negative, and even fierce and terrible characteristics:

> At the shores of the great earth facing the sea there will be nine foolish Shu. Sickness will be released (in the host family), among them fever, infected sores, infected intestines, mouth disease, eye disease, crippling deformity, deafness. Dust and the like will gather in eyes, decay in teeth. When walking among trees, branches will grab arms and (the walker) will be stuck between rocks (He Shicheng and Li Jingsheng, 1981a).

Before the 1950s, when Shu ceremonies took place every year in the second lunar month, Shu were offered medicine whereupon they would grant their people peace and progress in whatever they did. Often Shu would punish humankind, spreading disease or causing trouble among their livestock. Naxi believed that human beings must follow certain standards of behavior, including not cutting down too many trees, killing too many animals or digging too many ditches; to abandon such standards was to bring down the punishment of Shu on the community, livestock, and family members. An "altar for sending away bad things" was set up beside the altar of Shu during worship; before worship could begin, the malevolent Shu had to be driven away. In this way, the terrifying, mystic power of the Shu nature deities controlled human behavior.

In Dongba religious belief, two types of Shu exist: the benevolent and the malevolent. Both types sprung from an original formative feminine prototype, a chaotic, obscure unity of conflicting elements in an undifferentiated state. The nearness of Shu to feminine archetypes in their prototypical form presupposes that elements such as "benevolence" and "malevolence," far from being part of a clearly distinguished dualistic dichotomy, are still part of the "dual nature" of a greater totality.

The Transformative Character of Shu

The transformative character of the Shu feminine archetype is manifested during the ceremonies involved in Shu worship. One fundamental difference between these rituals and other Dongba ritual activities is that taking life is forbidden; no blood should stain the place where Shu worship ceremonies take place. Another distinctive feature is the presence of goat milk, poured into a bowl and offered at the altar. Both blood and milk are closely connected with femininity, particularly physical change in the feminine body in as much as essential physical changes are manifested by blood and milk. The first major feminine change, the coming of womanhood, is heralded by the presence of blood, while the second transformation, pregnancy, is caused by the lack of blood. The third transformation is the appearance of milk in a woman's body after childbirth. Just as blood in feminine transformation is concealed, a part of the transformation processes that is hidden and kept secret, so blood as a part of Shu as a feminine archetype must be kept from sight. The taboo against blood and life-taking at the Shu's altar is thus indicative of Shu's hidden, private aspect. Conversely, milk is representative of the creative processes involved in the birth of life and nurturing of life as well; it is thus a matter of feminine pride and public honor. It would appear then, that blood and milk are both associated with Shu as manifestations of the deity as a feminine archetype.

According to a Naxi myth, a Shu king (that is, a Shu) lived in the Meilidaji Sea (Lake) below Mount Naruoluo, while humans lived on the earth. One day, humans threw two copper stones that were red from the fire into the lake, causing the Shu king to become furious. Three days later, he punished humanity by causing the lake to rise up and flood the earth. Humans asked *Dongba Shiluo* to settle the dispute, whereupon he sent a holy Peng bird to carry the Shu Dragon King in his beak, flying around Mount Naruoluo three times and asking him to make the waters recede. Afterwards, Shu and humans made a pact not to harm each other. By having the Peng fly around the lake three times with the Shu in its beak, humans caused harm to the snakes and frogs within it. So humans must give medicine to Shu during Shu worshipping ceremonies (Li Jingsheng, 1991: 4). The myth however is more than explanation for the ritual act of giving medicine to Shu. While circling Mount Naruoluo, the Peng caused the Shu's body to become longer, showing that Shu are capable of such transformation. Milk, medicine and transformation are associated in Shu worship:

> Use the milk of white *bian* oxen, white *mao* oxen, white cows and bulls, white goats and sheep to make medicine. Mix in it 'Buddha's Palm' (a medical herb) and *ahluang*. Add plants that bloom in the summer and sleep in the winter. Use the warmth from silver, gold, *songlushi* and black jade. Use medicine to treat Shu and dragons, fish and snakes and frogs (He Kaixiang and Li Jingsheng, 1981).

Aside from the use of milk as a base for herbal medicine, what is worth mentioning here is that the herbs used are primarily those that "bloom in the summer and sleep in the winter," a specification that emphasizes the herbs' quality as plants that undergo transformative processes, changing shape with the changing seasons. Offerings to the deity are thus given the transformative character of the deity they are offered to, making herbs presented to Shu a direct reflection of Shu as beings that are capable of transformation.

A Shu's perch or place of residence ultimately rests on the Great Earth, which is the mother of all things, being that which gives birth to all things. Trees are the center of all life, symbolic of the prosperity and growth (fertility) of the Great Earth, while life-filled tree leaves, and nuts and berries are indicative of the feminine, life-giving qualities of Shu. Before the 1950s, one particularly verdant tree could be found among the verdant forests in front of and behind the Naxi houses. This tree was the perch of a Shu; a basket was hung from it and every year during Shu worship, bamboo and birch tree branches, paper flags and wooden plaques with images of Shu painted on them were placed in the basket and hung from the tree. In addition to this, offerings of eggs, milk, oat, etc., were made. The tree was the place where a Shu was fed and was called *Shu Baxu*. The great tree hidden in the forest is the residence or dwelling place of a Shu, while the Shu in question can also be viewed as the tree's spirit.

A ceremony known as "awakening the Shu" is performed as part of the ritual of Shu worship and after Shu are "awakened" they receive offerings.

> Awaken the Shu that dwell in the fields, by the waters, in the ditches. Awaken the Shu that dwell in black stones, on plains, in the villages, in the five directions of east, south, west, north and center. Awaken the Shu dwelling in trees, on cliffs. Awaken all the Shu between heaven and earth (He Shicheng and Li Jingsheng, 1981c: 127).

The various manifestations of Shu are associated with natural phenomena—forests, streams and channels, stones, etc. But while Shu have many different names and are omnipresent, they share one single, archetypical form, that of a being with a frog's head and snakes' tails connected by a human body.

In their manifestations as the spirit of nature or the Great Earth, Shu symbolize motherhood. The earth is the mother from which all life is born. Human beings live on her surface. Ancient Naxi perceptions of this spirit of nature as a feminine archetype is resonant with a feeling of being reliant on nature; human beings' reliance on the female gender springs directly from its maternal role as provider of protection, of nurturing and a capacity for accommodation to infants without which they cannot survive, while, in the same sense, human beings cannot survive without relying on the same qualities from "Mother Nature." Thus, nature is regarded as a mother and placed high in human esteem, holy and sacrosanct, a position that implies a kind of unity with untouched nature, symbiotic and of mutual benefit.

The Relationship Between Shu's Elemental and Transformative Nature

The elemental nature and transformative characteristics of the Shu as a feminine archetype cannot be neatly separated from each other, but rather are intricately connected, interacting with and complementing each other on many levels at the same time. The elemental character of Shu infuses them with a stable, conservative quality, while in their transformative aspect they represent continuous motion and change; this dual nature is in itself both constant and dynamic, like the "ying yang" of the *Book of Changes*, which only represents prosperity and good fortune to the degree that its two elements remain active, interacting, and in a state of flux. Femininity as a maternal force implies orbicular fertility and creativity, pregnancy and birth, and also the qualities of nurture and protection that the life it produces needs to exist. Conversely, creativity and birth as part of the elemental character of femininity are constant and eternal. From the creative process of pregnancy to breast feeding, mother and child are bound to each other by their protector-protected relationship, inseparable, creating a very special kind of mutual emotive attachment. But if this stage were to be incapable of transformation, the child would not grow up, and the creative nature of femininity itself would be obstructed. Change thus becomes a necessary element needed to assure that the elemental feminine characteristic of creativity can be retained.

Looking at all this in terms of the growth of consciousness and self-awareness in an individual human being, the unborn child still in its mother's womb and the infant during breast feeding both exist in a state where the unconscious world dominates, a stage dominated by the prototypical feminine archetype, that is by "elemental characteristics." As the child grows older, its sense of self-awareness increases, and "transformative characteristics" dominate. These two sets of characteristics, or aspects of a dual character, however, coexist in a complex relationship that is both symbiotic and conflicting, shifting and changing, forming and reforming, and thus maintaining their dynamic role in the dual character of femininity.

Neumann called this phenomena "psycho-gravitational dynamics," and explained it as:

> Psycho-gravitational phenomena cause specific unconscious elements to retain their unconscious nature while specific conscious elements are transformed into an unconscious, permanent laziness; in the meantime, within the relationship of the unconscious with the conscious mind, it interacts in conjunction with a special, predominantly feminine symbolism, forming the basis of the 'feminine elemental character' (1963: 27).

In the symbolism of Shu-related mythology, when the elemental character of Shu is ascendant, the Shu as a female archetype is manifested as a benevolent spirit or deity, while when its transformative character is ascendant it becomes malevolent.

Much of Naxi ritual involving Shu worship concerns asking for birth/fertility and seeking the creative power brought by water. At the same time, however, ritual is also informed by awareness of the transformative character of the Shu. Humanity naturally seeks both productivity and an affirmation of the processes of daily life and in so doing forces Shu to change their primary benevolent nature and reveal the malevolent side to their character. They restore Shu to benevolence through medical treatment and by making offerings. The annual rituals thus form the function of coordinating the elemental and transformative elements within the character of Shu.

Aside from the myths that tell us of the birth of Shu from eggs, others exist that describe them as half brothers. The society and culture of the Naxi's ancient ancestors was characterized as a whole by unconscious matriarchy, a stage during which feminine archetypes took precedence and Shu were predominantly feminine with masculine secondary characteristics. As time passed, patriarchy surpassed matriarchy and a new self-awareness separated itself from the world of the unconscious. Under these circumstances, the predominance of the feminine archetype in the construction of Shu began to be replaced by that of the masculine archetype. Thus, Shu were reconceived as half brothers, changing their gender.

Shu in their dominant traditional form represent a feminine archetype that includes both feminine (dominant) and masculine (complementary) traits. Not only does the structure of the most commonly used pictograph for Shu reveal this, but the rounded form of the total graph is also indicative of Shu's predominantly feminine nature, a "rounded, maternal femininity" (elemental characteristics of a feminine archetype) that is emphasized by the presentation of eggs and other offerings. Comparing the rituals involved in Shu worship with other Dongba ritual, we are struck by the roles of blood and milk as symbols of transformative femininity. The elemental and transformative characteristics of Shu as a feminine archetype interact with considerable complexity, at once contradictory and complementary, manifestations of the development of the feminine archetype itself.

REFERENCES

Fang Guoyu and He Zhiwu (1981) 'Naxi Pictographs,' in *A Complete Annotated Translation of Naxi Dongba Texts*, Yunnan People's Publishing House, Lijiang, Yunnan, China, First edition.
He Baolin (1981) Volume 18, Book 1, in *A Complete Annotated Translation of Naxi Dongba Texts*, Yunnan People's Publishing House, Lijiang, Yunnan, China.
He Kaixiang and Li Jingsheng (1981) 'Shu Worship: The Story of Cuidui Sanzi,' in *A Complete Annotated Translation of Naxi Dongba Texts*, Volume 7, Book 2, Yunnan People's Publishing House, Lijiang, Yunnan, China.
He Shicheng and Li Jingsheng (1981a) 'Shu Worship, Ritual Background,' in *A Complete Annotated Translation of Naxi Dongba Texts*, Volume 5, Book 1, Yunnan People's Publishing House, Lijiang, Yunnan, China.

He Shicheng and Li Jingsheng (1981b) 'Shu Worship: The Origins of Shu,' in *A Complete Annotated Translation of Naxi Dongba Texts*, Volume 5, Book 5, Yunnan People's Publishing House, Lijiang, Yunnan, China.

—— (1981c) 'Shu Worship, Asking Shu to Relax, Waking Shu,' in *A Complete Annotated Translation of Naxi Dongba Texts*, Volume 5, Book 3, Yunnan People's Publishing House, Lijiang, Yunnan, China.

He Yuncai and He Pingzheng (1981) 'Expelling Inauspicious Things, Story of the Ha and Si,' in *A Complete Annotated Translation of Naxi Dongba Texts*, Volume 36, Book 4, Yunnan People's Publishing House, Lijiang, Yunnan, China.

He Yuncai and Li Jingsheng (1981) 'Shu Worship: Story of Dusha-aotu,' in *A Complete Annotated Translation of Naxi Dongba Texts*, Volume 6, Book 8, Yunnan People's Publishing House, Lijiang, Yunnan, China.

Li Jingsheng (1991) 'Shu Worship Rituals and their Function,' in Guo Dalie and Yang Shiguang (eds.), *Dongba Culture*, Yunnan People's Publishing House, Lijiang, Yunnan, China.

Naxi Dongba Texts (1981) *A Complete Annotated Translation of Naxi Dongba Texts*, Yunnan People's Publishing House, Lijiang, Yunnan, China.

Neumann, Erich (1963) *The Mother Goddess: An Analysis of Archetype*, Translated by Li Yihong, Eastern Publishing House, Lijiang, Yunnan, China. Originally published in English as Erich Neumann, *The Great Mother: An Analysis of the Archetype*, Princeton University Press.

IV

Appropriation of Women's Indigenous Knowledge: The Case of the Matrilineal Lua in Northern Thailand

CHOLTHIRA SATYAWADHANA

This research is an investigation into the origins of the Lua people of northern Thailand, their matrilineal clan system, the narrative of how Lua women's knowledge of salt extraction was appropriated by outsiders, and the Lua confrontation with the forest conservation policy of the Thai government in the 1990s and particularly in the year 2000. This chapter is based on field investigations conducted over the last two decades in two areas:

1. the Thung Chang Pua Lua of Nan province where I did intensive fieldwork in 1978–82, 1986, and 1988;
2. the Doy Luang Lua of Chiang Rai who were forced to relocate to Lampang province where I conducted fieldwork in 1999–2000.

These two Lua groups share a cultural heritage of matrilineal spirit cults and are similar in many ways. Some subgroups of Nan in the former "red areas" gave up their matrilineal spirit cults when they joined the communist insurrection in the early 1960s. However, remnants of this belief system, social structure, and practices are still to be found among some other Nan subgroups in "pink areas." Field investigation in 1988 revealed that all Lua groups of Nan are the direct descendants of the Lua matrilineal social structure of Lanna of the past (Cholthira, 1991). The Lua of Nan province (a subgroup of Lawa) are the third most numerous among the highland peoples of Thailand who mostly live on the mountains of the Thai–Laos border. There are 146 Lua communities with a total population of 28,516, comprising 51.7 percent of highlanders in Nan province (Cholthira, 1987).

Anthropologists, particularly Paul Cohen, Peter Hinton, and Gehan Wijeyewardene (Cohen, 1984), have reported that the Lua of northern Thailand, especially those of Chiang Mai, practiced matrilineal spirit cults. One Lua group from Chiang Rai, who are now relocated in Lampang, believes that their ancestors settled in Chiang Rai before any other ethnic group and even before the Thai communities. Therefore, it may be inferred that the Lua of Chiang Rai are also the direct heirs of the forest Lua who lived in the early Buddhist Era, as mentioned in many ancient palm-leaf texts of Lanna (Cholthira, 1991).

The so-called northern Thai have been the subject of extensive literature about contemporary spirit cults in the region of the former Lanna Kingdom. Anthropologists assert that these spirit cults are organized around matrilineal descent groups and provide a venue for leadership and social control by women (Cohen, 1984; Davis, 1984). Much of the "Thai" and Lua systems of belief are centered around spirits. In turn, these systems are invariably associated with various forms of social organization, social relations, relative power, and ritual performances. It may be argued that power might have shifted from one dominant structure (Lua women) to another (Muang [Tai] men). Nevertheless, the original collective identity persists without its ethnic labels. Taking the Boe Kluea spirit cults as a case study, although "Luaness" and "Thainess" have intermingled both socially and culturally, separate Lua identities are still distinguishable.

The most ancient written source about the Lua is the inscription on the base of the Siva image at Kamphaeng Phet, Caruek Thaan Phra Isavara (1510) (Cholthira 1987, 1991: 136). This dates back to the 16th century and states: "According to the new law, sale of cattle to the Lua/Lawa was forbidden." It cannot be confirmed whether modern-day Lua are the same as the Lua/Lawa mentioned here who lived in the mountainous range and who had traded with lowlanders for a long period of time. However, a large number of cattle had been sold to the Lua probably for their carts and caravan trade, which was a major operation along the mountain ranges across the borders of northern highland principalities. Though we lack any accurate account, this caravan trade must have had a strong economic and political impact on Thai communities in the lowland, if the inscription reported the prohibition. This may be related to the reconstruction of Lua economic history. In the absence of accurate data, it may be surmised that the Lua economic situation in the past was not as bad as it is today. The importance of cattle and the Lua caravan trade together with Lua/Lawa archaeological sites show us a different dimension of Lua ethnohistory.

The area of Thung Chang Pua which the Lua inhabit today, particularly the Boe Kluea or salt mine area, was not always as remote as it is at present. More likely, it was at one time a strategic site through which every power "center"—Burma, Lanna, Lanchang, Chiang Rung (Sipsongpanna)—had to pass to dominate the extraction of rich natural resources, and the production of salt, both for everyday consumption and military purposes. There were a number of salt wells in Pua and the indigenous people, mostly Lua, had used them for generations. The salt from this strategic site was said to supply the needs of the whole of Lanna before 1950. During wars, these salt wells became more important for military strategy, as thousands of soldiers needed salt as well as rice. In 1450, historical evidence shows that the Boe Kluea (Lua) "state" was conquered by King Tilokaraja of Chiang Mai. Later, this strategic site was conquered by the Burmese. Power had shifted from one group to another during the Lanna crisis (1558–1774). In 1782, King Rama I ruled Bangkok and established Siamese power in Lanna. It was perhaps during this event that the Tai Lue of Sipsongpanna were forced to settle in the Lua area at Boe Kluea. Subsequently, McCarthy (1900) first marked the site "M.baw," at the beginning of the Nan river, as the location of salt wells with an enormous capacity to produce salt, noting that during that period, highlanders compared the price of salt to that of gold.

Lua Matrilineal Social Structure

Lua women have high status and significant roles in society and in their households. I interpret the Lua social structure as being matrilineal and matricentric. An old Lua saying (rendered here in Thai) points to the importance of women in Lua society: *ying raeng khwaai, chaai raeng kai* (Women have buffalo-like labor, men have only chicken-like labor). The existence of Lua matrilineal longhouse communities and their matricentric ideology confirms this interpretation. Yet, most ethnographic studies of the Lua of Nan before 1986 state that the Lua had neither clans nor lineages, much less matrilineal ones. David Filbeck (1971: Ch. 2, p. 1; Ch. 5, p. 6), in his research on the ethnography of the Lua, whom he called the "T'in tribe," first articulated this ethnographic interpretation in 1971:

> No clans or lineages exist among the T'in which would draw one's attention and concern away from his own village.... T'in society is village oriented with few or no strains of relationships running out to other villages. For the T'in tribal person, the village is the largest social unit.

Ten years later, William Y. Dessaint (1981: 128) also claimed that the T'in or the Lua of Nan had no unilineal social structure: "The descent system is bilateral, that is descent is reckoned both through the mother and through the father. There are no lineages, clans or other social institutions based on kinship apart from the family and the household."

In my first publication on the Lua, *Lua Muang Nan*, written in Thai (Cholthira, 1987), I mentioned Lua clans and lineages without, at that time, being aware of the fact that anthropologists Filebeck and Dessaint had denied their existence. The privileges among certain groups of particular ranks in the Lua hierarchy, coupled with the Lua terms *cao kok* and *traul* or *khra kul* (clan or lineage), are still observed in everyday life, strongly implying that not only do the Lua have clans and female lineages, but also that these matriclans are hierarchically interrelated. The mistaken inference that the Lua have a bilateral system is perhaps based upon observations among Lua who have abandoned their matrilineal traditions. In 1988, I went back to northern Thailand and traveled around various Lua communities in Nan to check my understanding of this contentious issue. This phase of research strongly confirmed that the Lua do really have matriclans or matrilineages. Living among the Lua for more than five years, I found that the life of the Lua was closely attached to water resources, which are in fact managed by clans. Lua people's daily meals include fish and other aquatic animals such as crabs, shrimps, and shellfish. Various aquatic plants are also favorite foods. In the Thung Chang Pua "red areas" (the base of the revolutionary Lua under the leadership of the Communist Party of Thailand [CPT] from 1967 to 1982), it is known that before "liberation," in some Lua communities, the privilege of using water and fish resources was organized by clans. There was a customary law which specified that members of other Lua communities, or those from different clans, were not allowed to poach on these resources, rights over which were handed down from generation to generation. Breaches could lead to serious conflict (ibid.: 47).

These ancestral rights over resources usually belonged to a major clan whose senior member had rank or was a person of authority in the community or among a number of interrelated communities. Two persons who claimed to have had such rights in the old days were the former female chieftains (ma? Rong) of Ban Kuuchaat, and the ritual leader (moe phii) of Ban Namchai. The operation of clans and lineages may also be seen in the way privileges were allocated in the distribution of land. Luas in the same village, though living on the same mountain for many generations, did not have equal rights in the choice of land for cultivation. Even nowadays, the clan of the chieftain, who is also known by the term cao kok (lord of the clan or descent group), which is the biggest clan of the village, has first right over choice of land for swiddening. This family also has priority in the recruitment of labor. Members of other clans cannot begin cultivation until the chieftain's schedule is completed (Cholthira, 1987).

The privileged clans of the Lua community, as I have described elsewhere (ibid.: 48), may be ranked as:

1. The big matrilineal clan, usually an extended family, which owned more than one rice field and more than 10 gourds. This matrilineal clan had privileges over fish and water resources, but they had to work in the chieftain's fields under a corvée system.
2. The ritual leader's matrilineal clan which had privileges over fish and water resources.
3. The super-matrilineal clan, usually the clan of the chieftain, which had the most privileges of choice over land, fish, and water resources, and was entitled to labor corvée from all adult members of every other matrilineal clan living in the Lua community.

I identified 33 matrilineal lines, with three types of identification which have been tabulated in Table 4.1. An investigation of the genealogy of the founding longhouse (*hüan kaw*) of one village, the Huay Thôn community, can be taken as a case study of Lua matriclans

Table 4.1
Lua Matrilineal Clans, Nan Province, 1988

No.	Type	Matriclan Names	Remarks
1	Ban	Ban Châ	*Ban* is a Thai word meaning village.
2	Ban	Ban Khom	
3	Ban	Ban Koak	
4	Ngual	Ngual Kaprual	*Ngual* is the Lua term for "village."
5	Ngual	Ngual Dakthiat	Also known as A? Vâl Dakthiat.
6	A?	A? Kal	Lua understand the word "A?" only as a prefix to clan names. No other meaning is given.
7	A?	A? Koa	
8	A?	A? Khat	Nos 8, 9, 10 may be the same clan, but their names are distinct in different villages.
9	A?	A? Khâ	
10	A?	A? Khân	
11	A?	A? Khin	
12	A?	A? Khoak	
13	A?	A? Khelr	
14	A?	A? Bo?	Nos 14 and 15 may be the same clan.
15	A?	A? Boang	
16	A?	A? Lol	
17	A?	A? Lab	
18	A?	A? Sâl	
19	A?	A? Sangkhâl	
20	A?	A? Sabung	
21	A?	A? Saweng	
22	A?	A? Siat	
23	A?	A? Sêk	
24	A?	A? Sapâl	
25	A?	A? Pih	Its origin was in Laos.
26	A?	A? Pelr	
27	A?	A? Pyeu	
28	A?	A? Dâr	Also known as An Dâr.
29	A?	A? Noang	
30	A?	A? Tdjûlr	Also known as Ban Tdjûn.
31	A?	A? Yelr	
32	A?	A? Yât	
33	A?	A? Yo?	Huay Thôn community's original clan.

and their function. In this original longhouse, every member worships the house-spirits, who seem to also be their matrilineal ancestral spirits. The Lua term is *prong/pyong tdjeng*, meaning "spirits of the house." In everyday life, *prong/pyong* has the same meaning as the term for ritual leader.

According to my field investigation, the name of the female lineage house-spirits of the *hüan kaw* is A? Yo?, representing the clan name of all members who actually live in it. It is evident that membership in the A? Yo? clan descends through the female line, that is, from (supposedly) the first generation to her daughter of the second generation, then to her daughter in the third, and similarly to the fourth generation and her daughter in the fifth. A male member is also an A? Yo? member at birth, but his membership is lost as soon as he is married. From then on he is no longer A? Yo? but belongs to his wife's female spirit line. In the A? Yo? longhouse, there was a man from another longhouse in the same community, but of another clan, A? Saweng. When he married Mae Coang Phat, who is an A? Yo? clan woman, he became A? Yo?. This couple had 12 children, of whom four died, while eight, four daughters and four sons, are still alive. Their four sons married into other clans or, as it is sometimes said, "to other spirits"; one of them is known as A? Pelr. Their four daughters were also married to men who came from other "spirits." One of them, who came from a Lua village in Laos, is known to have been A? Pih. All four husbands are now A? Yo?. Their children, both daughters and sons, are A? Yo?. Their grandchildren, the children of their daughters who all live in this longhouse, are alos A? Yo?.

There is another case of a man named? âw? û, now dead quite a long time, who formerly belonged to A? Pih in Laos. He also became A? Yo? when he married. His four sons, who were born A? Yo?, married and moved out to live in two neighboring longhouses in the same community. Two of them, including the youngest, are A? Khelr, while the other two are A? Saweng. Two of them now stay together in the same longhouse of A? Saweng, and the other two live in the same longhouse of A? Khelr. Both A? Khelr longhouse and A? Saweng longhouse are in the same community of Ban Huay Thôn where the original longhouse is A? Yo?. Their children are A? Khelr and A? Saweng, respectively.

From this genealogy, we may infer that there are at least five "female spirit" lines, namely, (*a*) A? Yo?; (*b*) A? Saweng; (*c*) A? Khelr; (*d*) A? Pelr; and (*e*) A? Pih (from Laos). Further genealogical investigation shows that two more lines occur in Ban Huay Thôn, that is, A? Lol and A? Pyeu. There are, thus, at present six "matriclans" for the five generations represented within the 11 longhouses of the Huay

Thôn community (since the ancestors of A? Pih lived in Laos, they never had their own longhouse in this community).

From this analysis of genealogy, we may understand that though there may be deviations in some Lua communities, especially among those who have moved to lowlands and live near lowlanders, the dominant system in Lua society is one of matrilineal clans, a system based on the Lua system of belief in female ancestral spirits and house spirits.

Women's Knowledge and Control of Salt Production

In the political economy of northern Thailand, salt production and trade played a key role in history. Although it has been argued otherwise, knowledge of salt production was clearly indigenous to Lua people, evident in their control of salt production and the myths surrounding its origins. Moreover, the dominant figure in Lua legends of salt mine origins was a woman, and the spirits of the salt mines worshiped to this day are all women, lending further credibility to the notion of matrilineal dominance among the Lua.

Dessaint's studies on the Thin/Mal or the Lua of Nan (Dessaint, 1973: 16) claimed that this Mon-Khmer speaking group controlled the salt trade throughout northern Thailand and Laos:

> Salt is collected by a simple technique in two communes near the head-waters of the Mae Nam Nan, namely Baw Kleue Nua and Baw Kleue Tay. Until recent years, salt, which is rare in the mountains of southern China and northern South-East Asia (there is still a high incidence of goiter in these areas), was used as an exchange standard. In the not very distant past, the 'Mal' living around the salt wells of Nan province controlled the salt trade by ox caravan or on elephant back throughout a large part of northern Thailand and northern Laos.

Later (1981: 122), Dessaint confirmed his view:

> Salt, as an exchange commodity, is a major source of cash or rice.... It was often used as an exchange standard. In the not very distant past, the T'in living around the salt wells in the northern part of Nan province largely controlled the salt production of the area. Lowlanders used to come to get salt in

exchange for rice or other commodities. They took it by zebu caravan or on elephant back to other parts of northern Thailand and northern Laos. This was a slow method of transport—ox caravans went from Muang Nan to Bo Kleue Tai in twelve stages, that is in twelve days—but it was a very profitable activity for the lowlanders who organized and led these ox caravans until recent times. In Muang Nan, they sold it for several times what it had cost them in Bo Kleue Nua or Bo Kleue Tay.

Chusit (1981) provided some supportive data for Dessaint's claim. His research on the ox-caravan trade of northern Thailand showed that the people of northern Thailand consumed salt coming from three directions before 1914: (a) sea salt from Malamaeng (Mataban?); (b) sea salt from Bangkok; and (c) hill salt from Boe Kluea of Pua district, Nan province.

During my own field research in 1986, I learned that the local people at Boe Kluea Tay believed that about 700 years ago, when the rulers of Phuu Khaa and Pua, Cao Luang Phuu Khaa and Cao Luang Pua respectively, were still good friends, they let their elephants walk and wander around the Pua area. After a month, they found that the elephants ate a substance which the Thai called *din poong* (literally, soil from salty pond). Cao Luang Pua tasted the water which the elephant caretaker (*khwaan chaang*) brought with him; it was salty. The Pua ruler then boiled the salty water, resulting in some perfect quality salt. From that time on, Cao Luang Pua invited Cao Luang Phuu Khaa to cook salt with him and the citizens of the two *muang* have used this salt mine since then (Cholthira, 1987: 190–91). If this legend is valid, knowledge of salt production was indigenous to this area, and not imported as suggested by some.

Legends often refer to the role of great ancestors or gods in the discovery of important skills or inventions: the art of spinning and weaving, the invention of sacred designs, the discovery of the most fundamental objects upon which a culture is founded, staple foodstuffs such as rice, domestic animals such as buffaloes, and basic raw materials for clothing such as cotton.

Yaa Lua and the Origin of Salt

Lua narrative refers to the role of a Lua grandmother (*yaa*) in the origin of salt and salt wells at Muang Pua. The legend of the origin of the salt mine among the Lua of Nan is:

Once upon a time, a long time ago indeed,
When people still lived together,
Having communal life,
Both production and consumption,
At that period the Lua did not know what salt was.

Later a grandmother, called Yaa Lua,
Took cooking as her duty for the Lua commune.
Whenever she cooked, the food was very delicious.
Everyone praised her and wanted to know
What kind of ingredients she added into her food.

Two men secretly watched her while cooking
And amazingly, they found out that, in fact,
The lady used the water she bathed in to cook food.
This secret was told and spread about,
But no one seemed to believe it.

So she was often watched
Until it was no longer a secret.
The people then held a meeting without her attendance
And it was decided by majority vote
That she should be punished by execution.

The Lua used a spear to kill her but she was able to run away.
Terribly hurt and bloody, she reached two wells:
Boe Nan and Boe Wen.
The blood flew down the well
Until the wells were full of bloody water.
These wells later became the salt wells of North Nan.

Grandma Lua continued to run away until she reached Boe Yuak.
She spat into the well, and
That is why the water in Boe Yuak is salty until now.
She continued to travel southward and died at Boe Luang,
Which is called Boe Kluea Tay nowadays.

(Story told to the author by Mae Phoeng, a Lua matriarch in Nan
in 1978.)

This legend was also used by the Lua to explain why the salty water
in Boe Wen and Boe Nan was red: because of Yaa Lua's blood. However, salt made from Boe Yuak and Boe Luang, where she died, was
white. The mythical Grandmother Lua died at Boe Luang, therefore

Boe Luang is believed to eat human beings. There had existed in the past, a tradition of human sacrifice every three or four years to appease the female spirit of Yaa Lua. Newborn babies to 10 year-old children were normally offered to the female spirit of the salt mine. Banana, sugarcane, rice, and meat were put in woven baskets called *krathong* and placed at the four corners of the well. If it was a human sacrifice, bits of human blood, heart, lung, and liver were also put in the *krathong* and offered. If it was necessary to seek a human being for sacrifice from a faraway land, only a bit of blood and the tip of an ear were needed (Cholthira, 1987: 191). Today, pigs, chickens, and dogs are sacrificed to Boe Luang instead.

Among the Lua in the mountain ranges, human sacrifice for the propitiation of the Boe Kluea spirit had been practiced consistently. However, it was given up with communist control of the area in the 1960s. A senior Lua lady told me frankly that outside the revolutionary base, the practice still occurred. Once she came across an outsider who seemed to be a ritual expert's follower, and she believes he came to steal babies. She threw a piece of firewood at the man, causing him injury and driving him away.

While the Lua myth and its associated sacrificial rituals record, on the one hand, the indigenous knowledge of salt production, on the other hand, the myth could also be a reproduction of matrilineal ideology. This "subjugated popular knowledge" is by no means borrowed from outsiders or imported by new settlers, because the "matrilineal" Lua ideological processes are still at work. No one is allowed to take the sacred female *sarong* down from the spirit shrine. The female tubular skirts have been among their spirits' offerings until now. In fact, the property of Cao Som Paak Nam, the last spirit of Boe Kluea, is still well protected by the male ritual leader, or Khao Cham. He showed me the *khrueang khrua*, literally "domestic utensils," of Cao Som. They included two female tubular skirts (*sarong*), one piece of Chinese silk—a delicate fabric with a violet floral pattern, a red traditional shirt, a pillow, a spear, a sword, a betel container (*uk maak*), and a food tray (*khan took*). All these interesting objects have become sacred materials offered to the female spirits of the salt mine and are used in spirit propitiation at Boe Luang once a year. In a three-year cycle, chickens are sacrificed the first year, a pig in the second, and a buffalo in the third. In all, in every three-year cycle, a black male buffalo, a dog, a pig, 12 chickens, two pairs of huge candles, and eight packs of small candles are offered to the Boe Kluea spirits.

Children were earlier sacrificed by the Lua to please the Boe Kluea spirits. This was a frequent practice particularly in the far northern area, i.e., the Boe Kluea Nuea. Two salt wells; Boe Haan and Boe Ket Sawaa, were closed by the indigenous people to stop human sacrifice. Nowadays, though human sacrifice is no longer practiced, remnants of Lua traditions still exist in the ritual performances, when buffaloes are sacrificed and their horns hung in front of the ritual pavilion. Likewise, dog sacrifice is still practiced and the Lua of Hang Dong in Chiang Mai cut the dog's ear and tail (without killing it) and combine these parts with other offerings made to please the ancestral spirits of the female lineage.

It can be seen in the Lua matrilineal system, and in the history and myths surrounding salt production in Nan, that women were the creators of popular knowledge and occupied a central position in Lua society. While this matrilineal society was unable to prevent Lua women's knowledge of salt production and spiritual propitiation from being appropriated by outsiders, it does seem to have prevented the disappearance of Lua identity as a whole. The question then becomes: how has power in this territory been shifted or subverted away from the Lua of Nan through the course of history?

The State Rewriting History

Lua yea' hai, Tai het na (The Lua work swiddens, the Tai work paddy fields)—a well-known northern Thai proverb.[1]

This saying may be interpreted as representing the cultural hegemony exercised by the Tai over the Lua through the centuries. Throughout *"la longue durée"* (Braudel, 1958), the history of the region is unveiled, not by institutionalized authoritative written texts but by the study of the popular knowledge of salt production and management and the spirit cults involved. Braudel (ibid.: 27) was totally correct when he said: "We may find a history to be capable of traversing even greater distances, a history to be measured in centuries, i.e., the history of the long, even of the very long time span, of the longue durée."

There are, so to speak, transformational laws of social relations and power relations which may be discovered. This is in accordance with Marx's assertion that social relations appear to us as if they were relations between things; in fact, in depth there lay relations between human beings. The process of the historical power shift from the Lua (Mon-Khmer speaking group) to the Muang (Tai speaking group) shows that state political power had shifted from Lua to Tai, and gender power had also shifted from women to men. In view of these shifts, the economic base of salt production in Lua society demands an analysis of social, gender relations.

My analysis of Boe Kluea, territorial spirits, and popular knowledge of the region must thus be set in its historical context. The Boe Kluea Lua community is seen in relation to the whole complex of districts and towns surrounding it, as well as the communities of neighboring states and countries. By doing this, we come closer to the dialectic of time spans, which will achieve an explanation of Lua society in all its reality, both historical and contemporary.

Lua history is part and parcel of the politics of the Lua states that developed and declined in the Lanna region and further north. What distinguishes them from the rest of the regional population is the fact that their society is an organization of matriclans and has been remarkably conducive to the emergence of matriarchs. It may well be that such distinctive institutions became firmly entrenched among the Lua of Nan during prolonged periods when they eschewed alliances with outsiders and withdrew their support for Yuan centers of power, after the latter either became excessively exploitative or the Yuan declined with the advance of the modern Thai domination. It is possibly the case that matriliny and matriarchy developed as village institutions with the "devolution" of the traditional Lanna kingdom (Cholthira, 1990: 75–101).

The Lua of Nan, including those of the Boe Kluea salt mine, have defied the dominant lowland states in their effort to preserve their identity and culture. Oral tradition suggests that such resistance took physical form at least four times over the last 200 years. Rebellions included the Suek Maan, Kha Tdjae, and Phi Bun Lua revolts, and the communist Lua insurgency (Cholthira, 1991: Ch. 3 and Ch. 4, Appendix A). Lua assertion of collective identity was later expressed in terms of customs, beliefs, and spirit cults without ethnic labels. The Lua–Tai assimilation is the product of a long economic, sociological, and political history. One result of this process is that many residents of northern Thailand are unaware that the Lanna traditions, which they recognize and revere, originated in Lua culture, which they themselves deem to be "primitive" and "uncivilized." Such traditions include the belief in the city-pillar, Sao Indakhiila, homage to Lanna (Lawa) territorial spirits, Puu Sae?, Yaa Sae?, the Phii Mot-Phii Meng dancing rituals, and the matrilineal ancestral spirit cults among the northern local people who now call themselves Khon Muang.

The spirit cults at the Boe Kluea salt mine of Nan province are also a strong case for Lua regional predominance in the past. Phii Muang spirit cults were held by the local people of Nan in the fifth month of every year, when 36 Phii Muang, literally city-spirits, were said to be possessed by spirit mediums, mostly female. If this information is correct, it suggests that there may have been at least 36 state rulers known in the realm of folk memory with a state history well preserved by folk wisdom in terms of spirit cults. It is suggested that the Boe Kluea region or Muang Boe (McCarthy, 1900) was first occupied by the Lua, and that a Lua traditional state emerged because of salt.

Recent history of the Lua of Nan may be explained in a broader framework of *la longue durée* or historical process and change. To do this we need to look at the evidence on Lua millenarian movements and resistance. The Phi Bun Lua revolt, the last Lua movement, is most prominent in their memory. Many young adults could recount their experiences in the movement. The rituals concerned suggest that the ideology was quite distinct from the Buddhist beliefs and was in fact truly indigenous. The chief of the movement Puu Wong offered, as sacrifice to the Lua spirit? eng Prong, a large pig five spans tall. Also, there was a ceremony known as *suu khwan* calling for an increase of spiritual power for the body, retained during the seven months that the movement lasted. Chickens were killed and sacrificed in the initiation ceremony for the new believers. Red strings were tied around the wrists of believers and it was proclaimed that there would be seven years of plenty when all Lua would live in prosperity. The messiah, Cao Phuu Bun, was coming shortly to bring the millennium to the people. He would bring with him modern goods and technology, turn each bamboo cottage into a cement building, and finally establish the Lua communities as a civilized state. On a certain day, Cao Phuu Bun would arrive in an airplane at Huay Khi Min village; indeed, it was claimed that he was then building the airplane.

While Puu Wong died in the Nan Provincial Penitentiary, his sons and sons-in-law were later released (Filbeck, 1971: 26). His wife, Mae Phong, who had also joined the communist insurgents operating in the area, is still alive and is highly respected by the Lua. When an ideological crisis occurred and the Communist Party of Thailand (CPT) found it difficult to control the red area, it was Mae Phong who made the decision and signed a treaty with the Third Army that the Lua people of south Nan would remain in peace. The Lua followed her advice and the military moved into the area without any bloodshed.

It is clear that the Lua have shared at least a united consciousness of rebellion with Mon-Khmer speakers in the state periphery. This ethnic consciousness was not all the "surface" of social relations; it was part of their collective consciousness, reproduced through the dynamic process of social history. The chain of historical events gathered from Lua oral tradition suggests that their ideology functioned as an internal and necessary component of the relations of production.

The Lua and the Khmu have engaged in a united struggle for quite a long time. Though each ethnic group has had its own version of "millenarian movement," differentiated in time and space, there was in particular one which was a kind of messianic movement active among some of the Mon-Khmer speaking highland groups of the areas straddling the borders within the periphery. The messiah who was to come and rescue the Kha, including the Lua (T'in) and Khmu?, and/or Lao Theung, all Mon-Khmer speaking groups, from their life of poverty, cultural disintegration, and economic deterioration was Tdjeueng. Amazingly, not only the Mon-Khmer speaking groups in Southeast Asia, but also the Hmong and the Yao within the periphery, including all Tai speaking groups, seemed not only to be involved with similar movements, but also shared the same figure as savior, Cüang.

The Lua were thus involved in the latest millenarian movement, the Phi Bun Lua revolt which took place in late 1964 and continued into 1965. This movement had close connections with the millenarian movement in Laos. After the unsatisfactory outcome of the movement, most of the Lua turned to the CPT and the region was known as a "red area." Oral tradition shows that the Mon-Khmer speaking groups of northern Thailand had been involved in a long chain of millenarian struggles. Such movements were also recorded in the written historical sources of northern Thailand in the series of chronicles of Muang Chiang Rung, Muang Lai, Muang Thaeng, and Muang Chiang Khaeng. In these chronicles, the Royal Siamese Government refers to the movements as Khabot KhaCɛ? (revolt of the KhaCɛ?), in some contexts as keut cɛ? pen Cüang. We may refer to it as the Kha Cɛ? revolt. Cɛ? (as represented in Thai in the chronicles) is a word used by Mon-Khmer speaking peoples; the Lua pronunciation may be represented as Tdjɛ?. The events first took place in the year 1861 and ended in 1884 when the Kha Tdjɛ? local troops were finally defeated by the modern military forces of King Rama V of Central

Siam. During these 23 years, a series of revolts took place in this Thai–Yunnan periphery. First, it was the Hmong who sparked the uprisings of other local people in the region, particularly the Kha Tdje?; that is, the Mon-Khmer speaking groups. Chronologically, the most important were the Sük Ho Thong Dam (Black Flag Ho) revolt in 1861, and the Sük Ho Thong Lüang (Yellow Flag Ho) revolt in 1862.

Chronicles provide evidence of the harsh taxation and conscription for warfare imposed on the people of Muang Thaeng, Muang Lai, Chiang Rung, and Chiang Khaeng. Possibly, the matrilineal structure of the Lua of Nan was a response, which allowed them to survive these hardships. Women's labor was crucial in agriculture as well as in the household, with men being lost in war and through corvée.

The Lua of Nan, who have been engaged in millenarian movements at least three times in the past 150 years, are illustrative of the Mon-Khmer speaking groups of the region. Their matrilineal social structure, in my view, is not a primitive stage in an internal unilinear evolutionary process. It may, in fact, be seen as "devolution"—an ancient Mon-Khmer kingdom being reduced to communities of a "primitive" and "stateless" form in which women need to hold power to maintain the processes of reproduction as well as agriculture and forest management. Matriliny, matrifocality, and matricentricity are constrained in a devolutionary process created by both internal economic conditions and external political power. Practicing "matriliny" in a strong sense among the Lua of Nan is perhaps the only way for them to preserve Lua identity.

Confronting the State Takeover

Thailand's rapid economic development has resulted in an equally rapid and degrading ecological transformation. It has also affected the resource base on which rural people's livelihoods depend. Although environmentalism in Thailand has grown considerably in recent years, environmental politics in Thailand are clearly more than a straightforward response to resource degradation. Environmentalism in Thailand not only reflects, but also acts upon, changing social and political relations at many levels.

In northern Thailand, in both the highlands and lowlands, including Nan–Chiang Rai–Lampang and their periphery, the mode of technocratic environmentalism practiced and politicized by the state via the forestry division has been the mainstream. Justification of state takeover has brought with it issues of serious conflict between highlanders and lowlanders. At present, the politics concerning environmental issues in Thailand is far from being a simple argument between those in favor of extremely green forest conservation and those in favor of sustainable usage and function of ecosystems. These arguments have initiated concerns among people's organizations and NGOs for the reform of the Thai constitution and the construction/reconstruction of a multifaceted civil society.

The Lua confrontation with the politics of environment played by the state via the forestry division is a case in point. The present situation of the Lua of Chiang Rai, who were forced to move from their homeland and were relocated in Lampang about six years ago (February 1994) according to a decision made by the Thai cabinet, is indeed a problem of community rights and state takeover. Along with other highland ethnic groups, the Lua of Chiang Rai have confronted problems of increasingly scarce agricultural land, resulting from the construction of hydroelectric dams and national parks. All these matters have affected the livelihood of the Lua in several ways, including their swidden cultivation and their rotated rice fields. This is Lua knowledge, which they have passed on from generation to generation.

According to an interview with the Doy Luang Lua at their new settlement in Lampang province (December 1999), all senior persons were born in Chiang Rai and some could trace their ancestral line back to over 10 generations. It may be hypothesized that the Doy Luang Lua or the Great Mountain Lua had settled in the Great Mountain of Chiang Rai for over 500 years according to their oral traditions, and for over 1,000 years according to Lawa palm-leaf texts. They had lived adjacent to a moderately fertile natural forest, which became a national forest park by a Thai cabinet resolution in the year 1993. In the past, the Doy Luang Lua community in Chiang Rai had relied heavily on forest products for their livelihood. They utilized forest products for food, and raw materials like bamboo for home use as well as for making handicrafts for sale. It is estimated that about 80 percent of their food came from the forest and homestead, as also medicinal herbs for traditional treatment, and leaves, flowers, branches, and tree-trunks as materials for ceremonial practices. Although Lua villagers earned a relatively low income, they lived peacefully and appeared to have good living conditions and good health. The Doy Luang Lua community also had a tremendous amount of knowledge concerning the utilization of biodiversities of animals and plants. Women and men differed in the possession of knowledge due to their different social and family roles and status. This knowledge, or, in other words, Lua wisdom, was generally found among women and men of older generations rather than younger ones.

However, Lua women had accumulated indigenous knowledge different from men. This is because Lua women were the ones who went out seeking food and firewood along the creek and in the forest. Children were educated along the Lua way of life led by the senior women. Therefore, it may be asserted that Lua women were potentially the key counterpart in ecological conservation in the Lua community. It has been the case that planning sustainable development is inefficient or likely to fail if women's roles and knowledge are overlooked.

Unfortunately, this invaluable local wisdom was destroyed and the process of learning interrupted by environmental politics when the settlement of the long-lived Lua community of Doy Luang was incorporated into the national sphere of influence. At present, the growth and strength of indigenous ethnic communities are usually considered a threat to national security. To lend legitimacy to the state takeover, the Lua community as well as many other hill-dwelling ethnic communities have become scapegoats for various social problems, such as poverty, resource conflicts, drug abuse, and crime. The Lua encountered a forced relocation of their communities, along with increased privatization and control of resources by the newly emerged state–market mafia.

In their new settlement at Ban Wangmai, in Wang Neua district, Lampang province, the process of state takeover and intervention has continuously weakened the Lua community's autonomy and its cultural identity. By making use of so-called "environmentalism," state policies on nature conservation influenced by western "extremely green policy" plus urban-based middle-class vision have protruded into the community's subsistence economy and its cultural space, particularly its traditional harmonious livelihood.

Although rural environmental movements led by some academic and NGO groups have supported indigenous highland communities in opposing state policies, especially on conserved areas, i.e., national parks, wildlife sanctuaries, and watershed protection forests, the situation of the Doy Luang Lua community at Lampang resettlement, as well as that of other ethnic communities such as the Lisu, Mien (Yao), Lahu (Musor), and Akha, which have also been forced to relocate in the adjacent areas, is still at stake at the moment.

Confronting state takeover and intervention, most highland ethnic communities in northern Thailand have become a social space of power struggles over natural resources and eco-politico-cultural domination. Community culture has been reconstructed to identify various forms of contestation between highland and lowland sectors at large. Reinvention of community-based forest management and a push for its recognition through the communal customary laws are, among other things, new trends in the indigenous communities' struggle for human rights, community rights, and land rights as well. In northern Thailand, according to Chusak Wittayapak (1999), these grassroots movements have currently widened to a "tribal-based movement struggling for citizenship rights and access to natural resources." These civic movements emerged in line with the emplacement of the new Thai constitution of 1997, which provides for community rights over natural resources as well as allowing people's participation in resource management.

It is not exaggerating to claim that the politics concerning environmental issues in the Thai state can also be interpreted as an obvious case of racial oppression and ethnic discrimination between the lowlanders and highlanders, urban and rural communities, and Thai and indigenous ethnic groups, where lowland-based state authorities have increased their efforts to oppress hill-dwelling ethnic minorities. The aforementioned case of the Boe Kluea Lua in Nan province in the historical scenario of northern Thailand fits well into this category. Recently, racist patterns and processes in the northern region have been augmented and transformed through acts of the state taking over land and forest resources.

In this research, I first drew out the past condition of land and forest use in one of the local Lua communities, the Boe Kluea of Nan province, presenting the coexistence of different and sometimes conflicting power relations, and the power shift from one ethnic group to another, which is related to and affects gender relations. At the same time, there is also a discourse of the "community" and a conscious effort to maintain communal rights to land and forest resources—not only salt but also natural resources—particularly the forest, in the present-day situation.

Through this contrast between "forest conservation," which is a matter of "state takeover," and "communal tradition" which is part and parcel of "community rights," creating a discourse of some of the regionwide cases and the confusion that exists in a specific locality, my point is not that the current discourse is merely a created tradition, but that such creative discourse and movement have sprung right out of a complex state such as Thailand—a plural society in its existence and reality, but a unified Royal Siamese Kingdom in its national, even racist, ideology.

In the final analysis, the Lua case evidently shows that some theories, related to a stereotypical paradigm of thought, were not based on reality, and that the indigenous groups have proved themselves to be good citizens, by participating in a process of formulating a "civil society," making use of their cultural heritage, knowledge, and wisdom to challenge the "powers-that-be" in order to preserve their identities and community rights.

NOTE

1. Tai is a linguistic designation referring to speakers of "Tai" languages in mainland Southeast Asia. Thai refers to citizens of the nation of Thailand.

REFERENCES

Braudel, Ferdinand (1958 [reprint 1980]) *On History*, translated by Sarah Matthews, Weidenfeld and Nicolson, London.

Cholthira Satyawadhana (1987) *Lua Muang Nan* (The Lua of Nan), Muang Boran Publishing House, Bangkok.

―――― (1990) 'A Comparative Study of Structure and Contradiction of the Austro-asiastic System in the Thai-Yunnan Periphery,' in Gehan Wijeyewardene (ed.), *Ethnic Groups across National Boundaries in Mainland Southeast Asia*, Institute of Southeast Asian Studies, Singapore, pp. 74–101.

―――― (1991) 'The Dispossessed: An Anthropological Reconstruction of Lawa Ethno-history in the Light of their Relationship with the Tai,' Ph.D. thesis, Australian National University, Canberra, unpulished.

Chusak Wittayapak (1999) 'Community Culture Revisited: Community as a Political Space for Struggles over Natural Resources and Cultural Meaning,' Paper presented at the 7th International Conference on Thai Studies, University of Amsterdam, July 1999.

Chusit Chuchart (1981) '*Phoe Khaa Wua Taang: Phuu bukbeuk kaan khaakhaay nai muubaan phaak nuea khoeng prathetthai* (Oxen caravan trade in northern Thai villages) (2398–2503), Ministry of Education, Bangkok, unpublished.

Cohen, Paul (1984) 'Are the Spirit Cults of Northern Thailand Descent Groups?' *Mankind* (Special Issue 3 on 'Spirit Cults and the Position of Women in Northern Thailand,' edited by P. Cohen and G. Wijeyewardene) *14(4)*.

Davis, Richard (1984) *Muang Metaphysics*, Pandora, Bangkok.

Dessaint, William Y.A. (1973) 'The Mal of Thailand and Laos,' *Bulletin of the International Committee on Urgent Anthropological and Ethnological Research*, No. 15, Vienna, pp. 9–25.

———— (1981) 'The T'in (Mal): Dry Rice Cultivators of Northern Thailand and Northern Laos,' *Journal of Siam Society*, 69(1 and 2): 107–37.

Filbeck, David (1971) *'T'in: A historical study,'* Ph.D. dissertation, Indiana University, unpublished.

McCarthy, James F. (1900 [reprint 1994]) *Surveying and Exploring in Siam*, White Lotus, Bangkok.

V

Gender Relations and Witches Among the Indigenous Communities of Jharkhand, India

SAMAR BOSU MULLICK

All the indigenous communities in mainland India suffer from gender inequality. As they are currently at different levels of social disintegration, patriarchy is also found at different stages of growth in them. Though the impact of external forces has been tremendous, the internal factors behind this growth are not any less active, however slow the process may be. This is true not only in the case of those communities who are settled agriculturists and partially practice foraging, but also in the case of nomadic foraging ones. Even the matrilineal Khasi and the Garo register increasing dominance of their men over the communal resources of which their women are supposed to be the custodians. These communities are not, however, at the same level of social change in terms of gender relations among them.

The socioeconomic factors responsible for such changes are active in varying degrees of intensity, facing equally uneven degrees of opposition from the remnants of the ancient forms of relations of production and their related value system. However, the impact of the external forces on this process of change is becoming increasingly powerful. But the transformation of a gender-just society into a completely patriarchal one is a radical social change, which requires centuries, if not millennia, depending on the internal growth of socioeconomic factors responsible for this.

The objective of this chapter is to see if there is any co-relation between the growth of inequality in gender relations and the widespread belief and practice of condemnation of women as witches, particularly among the Munda and the Ho in Jharkhand in northerncentral India. A comparative analysis has also been made between the economic and cultural conditions of the Ho and the Munda on the one hand, and other Kherowal communities on the other, in order to understand the relationship between the belief in witchcraft and degradation in gender relations. Two witch songs have also been studied in great depth to gain insights into the past and present gender relations in these societies.

The belief in the existence of witches and the efficacy of witchcraft is found to be deeply rooted in the belief systems of the patrilineal agriculturist communities of India. They, however, are remarkably absent among the matrilineal Khasi, the Pnar and the Garo, as well as among the nomadic foraging communities like the Birhor and the Erenga Munda in Jharkhand, which are patrilineal. This fact may lead us to presume that the belief in witchcraft has some cultural connections with the economy of settled agriculture, and the patrilineal social structure. We may even go further—into the realm of magic— to identify the fundamental reason behind the growth of the belief that "evil" can be more powerful than "good." Thus propitiation of "evil" is necessary to get rid of pain and death, as opposed to the practice of propitiation of "good" for protection against such odds. The reason, once identified, can then be used to explain the origin of witchcraft, and in a later stage of social development, its use to degrade women. The root of the general notion that witchcraft is the misapplication by individuals of magic, which was designed for the service of the community, actually goes much deeper into the social life of the community than just the moral or ethical level. The reason why only witches, not wizards, are punished with a high degree of cruelty has to have a wide social perspective. It is true that the dominant value of social life in the indigenous communities is collectivism, and anything that threatens it is branded as antisocial. But why are women considered more antisocial than men, who are never or seldom identified as witches? The question remains unanswered. This chapter endeavors to see the belief in witchcraft in the context of the changing socioeconomic condition of the indigenous peoples in India.

Origins of Witchcraft

Myths and Legends

The Munda and the Ho belong to the family of the speakers of the great Austro-Asiatic languages. Their languages are placed more precisely in the north Munda branch of the Munda subfamily. The closeness of their respective languages proves their sociocultural commonality. As far as the belief system is concerned, they in fact disagree with each other on nothing. Similarly, they share their social system and historical tradition. The only possible difference may be found in the degree of exposure to the surrounding dominant Hindu society.

Both the Ho and the Munda share the same myths and legends. As part of their oral tradition, these myths and legends were transmitted from generation to generation orally until the end of the last century when ethnologists started collecting them. They are now available in print in the two great ethnological works of this century—the *Encyclopaedia Mundarica* (1950) by Reverend John Baptist Hoffman, S.J. and Arthur van Emelen, S.J., and *The Mundas and their Country* (1912) by S.C. Roy.

The origin of witchcraft is narrated in the legend of *Baranda Bonga*. The legend, as translated and described by Hoffman, reveals how the first ever known witch in the collective memory of the people is identified by two non-Munda people. She is none other than the wife of the Supreme Being, the creator of the universe, *Singbonga* (literally, the supreme spirit). She is accused of killing *Singbonga's* son (obviously born out of his other wife). *Singbonga* asks his elder brother *Baranda* to get rid of her. *Baranda* carries out the order obediently but not before proving that she is really a witch. The legend presents the two witch-finders as possessing great magical power. They are introduced as the *guru* (master-teacher) and the *mantri* (minister) who are tilling land with a plough drawn by a pair of tigers harnessed by snakes. *Baranda* approaches them to find out the cause of his nephew's sickness, i.e., the only son of *Singbonga* (we find no other story where *Singbonga* has another son). Thus he plays an important role in beginning the process of witch-finding and in ending the life of the identified witch. The legend does not reveal whether the wife so identified is the first or the second wife of *Singbonga*. Usually in rural folktales, the king banishes or kills his first wife under the influence of his younger second wife and her associates. If this rule is followed, then the wife in the legend has to be the elder one and the son, the offspring of the younger one, and only then can the alibi of the murder be logically established, i.e., supplantation of the innocent "old" by the cruel "new." In other words, it is the condemnation of the older gender relationship and introduction of a new one with an inherent brutal inequality.

The legend introduces many new things into the Munda belief system. In the original myth of creation, the *Sirijan Durang*, the Supreme Being is self-created. There is no mention of his wife/wives, son or brother. In the destruction myth, the *Sengel da* (Rain of Fire), also these characters are not found. Only in the third important myth, "The Making of the First Plough," do we find the wife of the creator. The later version of the *Hasur* or *Asur Kani*, (the *Asur* legend, the only known ballad in the Mundari language), also introduces *Daibi Kumari* (a non-Mundari word) as the wife of the creator, but still no son or brother appears. Among the legends accumulated under the name *Baranda Bonga* or *Baranda Buru* (the *Baranda* spirit or *Baranda*, the mountain spirit) two are important. The first one introduces *Baranda*, a so far unknown spirit, as the elder brother of *Singbonga*. The son and the second wife appear in the second legend, which is also the last one in the sequence. Besides, it endeavors to present *Baranda Bonga* as the central figure and introduces the practice of witchcraft and witch-finding, which is said to be "something new and formerly unknown" to the Munda. However, the most striking feature of the legend is the very character of the Supreme Being. Here he is presented not as the familiar "old one," loving and compassionate, but as an ordinary human being. The legend, if seen in association with the ones mentioned here, reveals not only an attempt to introduce a new system of belief into the prevailing Munda faith, but also the transformations in the Munda belief system. We may identify three phases of such transformations.

1. In the earliest known system, there are only three characters in the pantheon, *Singbonga* (*Buru Bonga*), the Supreme Being, *jaer* (*jaher*) *era*, the matron of the sacred grove, and *Nage era*, the matron of water. When the Supreme Being, originally known as *Haram* (the old man) left the human beings that He

created, forever, He became the great mountain spirit (*Marang Buru*). There was no evil spirit to be propitiated. The presence of both *Singbonga* and *jaer era* in the Santhal system proves that they were the dominant spirits even before their division into two distinct groups, the Santhal and the Munda, from the original one. Similarly, at the ritual level the *Baa* (Munda-Ho) or *Baha* (Santhal) belong to the same period. The most remarkable fact that should be kept in mind from our perspective is that the Flower Feast is presently the only occasion when women exercise their right to ritually propitiate ancestral spirits. In most villages, which are not overwhelmingly exposed to alien influence, the role of the village priest (male) is almost negligible.

2.   In the second phase, we come across two important legends, "The Making of the First Plough" and "The Division of Time into Day and Night." In them, the "old one" teaches the "son of man" the technique of making a plough, and the art of tilling land. He creates the moon to carve night out of day so that the tiller can rest after the hard work of the day. The legend not only introduces a consort of the "old one," but also presents her as his "better half." It tells us how the Supreme Being who created everything fails to make a plough properly despite his best effort. Then his wife comes to his rescue and in no time constructs it using much better technology than her husband. The "old one" accepts "defeat" and declares, "From today I free all women from the hardship of making ploughs. The men shall make them and the women shall not even plough them."

3.   The third phase can be divided into two subphases: first, we have the *Asur Kani*. It introduces a set of female spirits, as the protectors of different parts of nature created by the Supreme Being. But they are to be propitiated by the male priest of the village. In the social plane, he declares the need to strike a balance between the need for generation of material goods, and the regeneration of physical nature. He shows the difference between economic necessity and selfish greed, and for him, life is more important than matter. Now the Supreme Being is definitely called *Singbonga*, one who lives up in the sky. An analysis of his description in the story along with a comparative study of the growth of the Supreme Being concept in the tradition of other groups belonging to the same indigenous community conclusively indicates his association with the great mountain first, and then the moon (Bosu Mullick, 1991). We may remember that earlier, he was called the sprit of the

great mountain, *Marang Buru*. In the second subphase, we find a number of legends and myths that gather around the new mountain spirit, the *Baranda*. They describe the gradual association of the Supreme Being with the sun, degradation of *Nage era*, the matron of water, into an insignificant tutelary spirit, and a grand introduction of an "elder brother of the Supreme Being," *Baranda Bonga*. Now the traditional practice of propitiation of "good" is unable to protect human beings from "evil." On the social plane, women are branded as the potential keepers of the evil spirits; to be killed if the newly introduced witch-finders identify them as witches.

Hoffman strongly believes that the Hindu religion influenced the Munda belief system with respect to the development of the institution of witchcraft. He makes two observations on this. His observation on the *Asur* legend is,

> The ritual details prescribed are all of them proper to the cult of *Singbonga*, and it is pretty evident that the whole legend is destined to show the victory of the Sun cult over a religious system previously prevalent.... It is difficult to think that the legend is a Mundari creation. Until reasons to the contrary are found, I incline to attribute it to those Aryan missionaries, who, in very remote times tried to convert the Aborigines to their religious views, and who presented their rites and doctrines in a form harmonising as far as possible with the monotheism they found among them (Hoffman, 1950: 243).

In the following passages, Hoffman explains Aryan missionaries as aborigines who were well-versed in Aryan religious doctrines and tried to introduce the sun cult ideas to their own communities (ibid.: 246). He notes,

> The *Asur* legend might be considered as an attempt to justify the transition of the Munda race from the patriarchical to the monarchical stage of monotheism. That the system which teaches the existence of evil powers in witchcraft and the power of its priests to counteract the evil, is a later introduction is implied as well as distinctly stated in the *Baranda* legends, where it is also expressly stated that the Mundas feel the introduction of the system to be regrettable (ibid.: 250).

About the use of the legend in the rites of *Soso Tapa* (performed annually to ward off the impact of the evil eye on the family), he thinks that since the rites consist essentially of protracted praise of *Singbonga* it must be originally intended to counteract only "poisonous" praises. The belief in its efficacy against the evil eye is also a later development (ibid.: 4073).

His second observation is on the legends of *Baranda Bonga*. He holds that the cult of *Baranda* is of non-Munda origin. It is a new system of religion, which was introduced to the Munda by the more Hinduized artisan groups, such as, *Barae*, the blacksmith, and the *Turi*, the basket maker (ibid.: 422, 425). Many Munda communities got Hinduized and subsequently submerged in the conquering Aryan society. They might have acquired the system there and passed it on to their brethren living up in the hilly forests. Elsewhere, however, he concludes,

... the cult of *Baranda bonga* and the practice of witchcraft have been borrowed from the *Sadan* (the Hindu migrants in the Munda country), and that one such separated branch of the Munda race has been responsible for their adoption by all the Munda (ibid.: 428).

In yet another place he says, "The cult of *Baranda* came from Hinduized aborigines, probably such as were of Dravidian origin: among these, the belief in distinctly evil spirits seems to be much stronger than among the Munda" (ibid.: 431). He repeats this view while writing on the witches separately.

It is this belief and practices connected with it (the witchcraft) that one of their historical tradition states, that it has been introduced into Chota Nagpur by the Oraons. And one of the *Baranda buru* legends states explicitly that the system is an innovation... (ibid.: *najom*: 2918).

He substantiates his observation that the system is of alien origin and of later introduction among the Munda with the following facts,

The ministers pretending to counteract the powers of wizards and witches, i.e., the *sokhas* and *deorans*, need neither belong to the Mundari village family nor even to the Munda race. The *sokhas* are mostly Sadans or Oraons. The language used by

them is mostly Sadani or Hindi and hardly ever understood by
the Mundas using these mantras. The texts given by Mundari
deonras are a corrupt Sadani, which is hardly understood even
by Hindu *pandits*. The source from which genuine *sokhas* and
marang deoras derive their powers (*gun*) is not *Singbonga*, but
Mahadeo (one of the Hindu trinity), it is not even clear whether
he be really a personification of the principle of good. The form
of worship offered to him, are purely Hindu rites (ibid.: *najom*:
2918–19).

Excepting a few Hindus most of the *sokhas* are Hinduised
Mundas. Near their house there is a small enclosure with a little
earthen platform on which a turusi plant (*Ocimum sanctum*) is
grown and a large round pebble in which *Mahadeo*, the divinity
they worship, is supposed to reside. The Hindu trident is also
planted there, and close by, a bamboo with a small triangular red
flag. There the *sokhas* make weekly the offerings proper to
Hindu worship: milk, flower, incense, red lead, etc. (ibid.:
sokhas: 4033).

These transitions on the level of ideas and institutions certainly
reflect Hinduization of Munda beliefs, and the changes in the material
life of the people. In considering the three phases in the legends discussed earlier, we find a brilliant picture of the growth of patriarchy
among the Munda and Ho people as a result of both internal developments, and external influences. It also delineates the economic transformation of the society from hunting-gathering to settled
agriculture.

Gender Relations and Mode of Production

Kelkar and Nathan (1993) have analyzed the rationale behind the
uneven growth of patriarchy among the indigenous communities of
Jharkhand. My own field observations support the same view. The
relations of production in these communities are at various levels of
their incompatibility with the corresponding forces of production. If
in the scale of gender relations, the foragers occupy the more just
side, the rest are found on the other end. Among the latter, if the
Santhal occupy the more just side, the Munda are on the other end,
while the Ho occupy a place in between. But in all these cases, the

forces of production have changed a lot leaving the relations of production that evolved in an earlier period of their history far behind. The Santhal are now fairly exposed to the stratified social system of the dominant Hindu society. They have no trace of their ancient communal ownership of the means of production, such as, the land and the forest. The relations of production, and the related customs and practices however have not changed at the same pace. They have still retained the rituals relating to the old practice of redistribution of land according to the need of the different families in a village, although in reality, it is not practiced. The same practice exists among the Munda and the Ho, but only in their collective memory.

The institution of the *Manki* (the head of the confederation of a group of villages numbering 12 or 13) of the Munda and the Ho can be compared to that of the *Pargana* of the Santhal. But the political power and the economic control over the ancestral communal land and forest that the *Manki* enjoyed during the mediaeval, feudal, and the colonial periods cannot be compared to that of the *Pargana*, who in all probabilities emerged as a political head of a cluster of villages among the Santhal in the feudal period. While the *Parganas* all along opposed the state aggression on the Santhal, a section of the *Manki* cooperated with the process of state formation, and during the colonial period even declared themselves as the *zamindars* of the communal land and forest. A section of the Munda who later became known as the "Bhumij," even went a step further to form a primary state in Barahabhum in the Jangal Mahal region (presently in West Bengal) (Sinha, 1961). The Ho, however, remained relatively isolated and kept opposing the aggression of the Porahat state into their territory. Their valor in the struggle against the British forces is well recorded. Thus, the *Manki* among them could never aspire for the status of a zamindar. The Munda of Mankipatti region (present day Khunti subdivision of Ranchi district) retained this trait for quite a long time as compared to those of the adjoining regions of Panchpargana (eastern part of Ranchi district and some portions of east Singhbhum district) and Naguridisum (Gumla district). In this region, which remained relatively isolated from the state system till the turn of the present century and known widely for the rebellion of Birsa Munda, the practice of holding land communally remained alive till the enactment of the Chota Nagpur Tenancy (CNT) Act (1908). The kind of degradation in gender relations and decrease in women's rights in land that is observed in Manbhum, Panchpargana, and even Naguridisum, is not found in the Mankipatti area. In fact, women's right to land among the Munda and the Ho of Chota Nagpur as enshrined in the CNT Act was modeled after what was found in this region.

The presence of and the right to access the forest in the region can to some extent explain this unevenness in gender relations. It is generally observed that in more forested areas, women enjoy more equal status with men even if they have less access to cultivable land. Since the possibility and practice of foraging is quite high in these areas, women are more economically independent and less dependent on agricultural produce. On the other hand, where there is no forest left or its presence is negligible, women are at the receiving end. High dependency on agriculture leads to an assertion of women's traditional rights on family land for independent survival. For this assertion, they incur their in-laws' or even brothers' wrath.

"Among the Hos," Dalton observed,

> all diseases in men and animals are attributed to one of two causes—the wrath of some evil spirit who has to be appeased, or the spell of some witch or sorcerer who should be destroyed or driven out of the land (Russel and Hiralal, 1969: 513).

The Santhal and the Munda share this belief, but the foraging communities subscribe to this only partially. They believe in the existence of evil spirits but not in witchcraft. It is interesting to note that neither the Kodaku nor the Hill Korwa practice witchcraft. But they have developed the institution of *dewar* (sounds similar to the *deonra* of the Munda and *dewa* of the Ho).

> *Dewar* is a shaman of the village by virtue of his acquired qualities through training given by a guru. The *dewar* may belong to the group or he may be a non-Korwa. If any one falls ill, the *dewar* is invited to name the spirit responsible for it (Singh, B., 1977: 4).

Among the Kodaku, "the *dewar*'s service is sought (after) and utilized at a time of any emergency or crisis like breakout (outbreak) of any epidemic, to appease the spirit who is believed to have caused such illness..." (Singh and Danda, 1986: 89). But the Birhor and the Erenga Munda of Saranda forest (Hoffman, 1950: 1315) have not registered any institution like the *dewa*, *dewar* or *deonra* though they believe in the presence of evil spirits.

Agriculture among the Hill Korwar is not as advanced and as dominant in their economy as it is among the Santhal, Munda and the Ho. The Kodaku are, however, complete foragers, though now some individuals work as agricultural laborers. Both the communities believe in the influence of evil forces (spirits) on human life. They also think that these evil spirits are to be propitiated separately. For this purpose, they have adopted the institution of the *dewar* from their Hindu neighbors, but they have not yet adopted the belief in witchcraft. This places them in a third category between the Santhal, Munda and Ho who believe in both existence of evil spirits and in the potentiality of some women as the keepers and controllers of such spirits. Thus, we find three stages of the development of belief among these people. First, belief in spirits, including the ancestral ones, who are benevolent. Both women and men possess the ritual knowledge to propitiate them. Second, an additional belief in evil spirits who are to be propitiated separately by diviners (*deonra, dewa* or *dewar*), an institution different from the village priest, the *pahan, naeke, diuri* or *baiga*. Third, the belief that human beings can become intimate with and control evil spirits. Some women and men often do so to cause harm or even kill their fellow villagers. They are the witches (the *najom, dain* or *dan*) and are to be identified and punished. Only some specialists who are trained in this profession by Hindu *gurus* can perform this job of identification. This is followed by a marked shift to the identification of only women as keepers of evil spirits.

These three stages of belief are found to be associated with three different stages of mode of production, namely: foraging; foraging-cum-agriculture (where foraging plays the dominant role and state of agriculture is rudimentary); and agriculture and foraging (where settled agriculture is practiced as the dominant productive activity).

The respective relations of production show the changing gender relations. These societies are divided only in terms of gender. However, if women and men are considered the first forms of class, then in these three stages we can see the gradual development of a class struggle which leads to the establishment of patriarchy through several stages of weakening of the female rights, both economic and ritual (Nathan, 1997). Among the indigenous communities of this study, male domination is an established fact. The variations show only different phases of transition of the society towards full-grown patriarchy, and the corresponding political formation of "state." Some sections of the Ho-Munda community, like the Bhumij and other small groups in Manbhum and eastern Ranchi plateau, have already reached the highest stage in this transition and participated in the process of state formation. Another section is still at a lower stage of this development, like the Erenga Munda and the Birhor. In the former case, the women have almost lost their traditional rights, and in the latter, the men are not economically as independent of them as to be able to curb their rights entirely. Thus in both the cases, the gender contradiction is not very sharp. However, it is very sharp in the case of groups where women still have a considerable share in the means of production, and where the level of agricultural productivity is considerably high, thereby enabling men to become economically independent of women.

Women played an important role in this process. According to the old custom, which is still practiced when the Munda go from one village to another, their wives lead them. Their knowledge of seeds is acknowledged and used even today. The knowledge of the roots of a particular plant used to brew rice beer, the most sacred as well as favorite drink of the people, is a secret known only to them. After marriage, the woman passes on this knowledge to her husband's family. Her knowledge of herbs and plants, particularly medicinal ones, is considered a precious family possession. Chota Nagpur proved to be a treasure house of such knowledge because it held a rich and wide variety of flora and fauna, which the women came across while foraging. Though the dense forest has now faded away, foraging is still a favored preoccupation. Their role in the preparation of cultivable land cannot be ignored either. Even today, wherever we see land being reclaimed from forest, women work together with men. Women's contribution to the development of agriculture is further confirmed by the "myth of the preparation of the first plough." Here the Supreme Being's wife is described as the real inventor of the technology of plough-making. Thus, the woman's right to the newly reclaimed land and its produce was naturally established and recognized by the community, which received a permanent place in their customary law.

Though the Santhal and Munda were located in different parts of Jharkhand, they developed a similar social system but with a difference. The difference is found in the development of two dissimilar sets of myths and legends, rites and rituals, dance and music, arts and artifacts, and so on. They faced different kinds of natural and social environments and were exposed to varying types of alien people and alien cultures. The new agricultural technology that they borrowed from their alien Hindu neighbors and the hitherto unknown state system that they were forced to interact with, left varying marks in their emerging social systems. These variations have naturally contributed to the uneven growth of patriarchy.

We find three types of societies in India in this context. First, in the Hindu caste societies, the belief in witchcraft is quite strong but witch-hunting is not practiced. The other nonindigenous religious groups like the Muslims, Christians, Buddhists, etc., fall within this category because an overwhelming majority of them are descendents of original Hindu lower caste converts. Belief in witchcraft is as strong in them as in the Hindu caste groups. Second, in indigenous societies, the belief in witchcraft plays a very important and vital role in the life of the people. The whole community is found to be engaged in an apparently never-ending orgy of witch-hunting. Third, there are small and partially settled or nomadic indigenous communities who have either no notion of the existence of witches, or are least bothered by them, such as the Birhor of Chota Nagpur and Korku of Sarguja in the Jharkhand region, and the Khasi and Garo of Meghalaya.

This unevenness in the belief in witchcraft needs explanation, but first let us examine the two words, *dan* and *dakan*, which stand for "witch" in Santhali and Bhilali respectively. Both words have their root in the old Indo-Aryan word, *dakini* (Chatterji, 1970: 308). Interestingly, the word witch comes from the Anglo-Saxon word "wicca," meaning "wise one" or "magician" (*World Book—Multimedia Encyclopedia, Witchcraft*). Similarly the original meaning of the word "*dakini*" can be traced in *Tantra*. According to Tantrik Buddhism, *dakini* means the female personification of a stage of wisdom (Rawson, 1978: 210). A popular image of the Hindu goddess Kali is found to be associated with many other images, such as, Shiva, her so-called husband, the fox, her "vehicle," and *Dakini* and *Yogini*, her two associates. The image of *Dakini* is that of a curved female goblin attending Kali. According to the *Pauranic* tradition, Kali is *Prakriti*, the "mother goddess," the symbol of female principle. Thus, if Kali is the deified form of the female principle, *Dakini* has to be the real form of the same, and she represents female supremacy on earth. In the Buddhist *Tantra*, Durga, the female deity of the Aryans, is called "*Vajradakini*." *Dakini Vidya* means witchcraft. So her knowledge is the knowledge of the witch. Her mate *Yogini* is a shamaness. She is also considered the personification of *Bhagawati* (Chattopadhyaya, 1959: 326). We know that with the changes in gender relations from mother's right to father's right, the deification of female power in the level of religion and denigration of it in the level of institutions take place in the new social order based on patriarchy.

It is to be noted at this point of our discourse that neither the Santhal nor the Bhil have words for witch in their own languages. This fact may lead one to believe that they borrowed the idea and the institution of witchcraft from their neighbors, most probably the Hindu castes. They did that at some point of their social evolution to meet certain social demands. At what point of time in their history this demand arose is a relevant question, which we will address later. Another important fact to note is that originally, a witch was either a woman or a man who supposedly had supernatural powers. Through the years, however, mostly the women who came to be considered as witches. Among the Munda and the Uraon in Jharkhand, both witches and wizards are still found. Among the Bhil, wizards are called *bhutala*. But neither among the Bhil nor among the Santhal are wizards found anymore. The point is: why and how did the meaning of the word witch, *dan* or *dakin*, reverse diametrically? What are the social factors responsible for its transformation from a word of respect and awe to a word of abuse and hatred? Moreover, what happened to the tradition of wizards? The songs under consideration hold the answer.

Interpreting Witch Songs

The phenomenon of "witch-hunting" is of great concern for both civil society and the state. Yet, the lack of understanding leads one to conclude that it is mere superstition. Therefore, the common prescription for its eradication is the "spread of education" among the people who practice this inhuman belief. The fact that even educated people are still deeply influenced by this belief negates the efficacy of the prescription. There is certainly an urgent need to make people aware of the truth behind this ancient belief, but mere spread of education cannot achieve it. It requires specialized research in the subject. It is quite strange that in India, apart from a few individual exceptions, neither anthropologists nor administrators have endeavored to delve into the matter to scientifically explain the phenomenon and find a solution. The available literature on witchcraft or witch-hunting is basically descriptive/narrative. They are, however, most valuable for they embody facts covering more than a century. *Descriptive Ethnology of Bengal* by E.T. Dalton (1872), *A Memoir of*

Central India by Sir John Malcolm (1880) and *The Tribes and Castes of Bengal* by H.H. Risley (1891) are the earliest works to inform us about it. Kelkar and Nathan (1991; 1993) have conducted pioneering work in analyzing the phenomenon in recent years. This has been the result of a growing consciousness of gender issues emerging out of the women's movement in the country. My search follows the same track with necessary modifications depending on the demand of the subject in focus.

It is believed that the witches dance and sing together on certain days, and that they have their own songs for these occasions. There are songs too, which are sung not by the witches but by their opponents. I call both these types of songs "witch songs." During my search for material on witchcraft, I came across two very fascinating and revealing songs. Both are obviously composed by the male members of the respective communities, namely, the Santhal and the Bhil—the *Kolla* and the *Bhilla* to use their ancient names. The Santhal, the second largest indigenous people's group in India, is the largest community of Kol. Some authors, however, believe that Bhil are a Dravidian tribe (Risley, 1891: 370). Yet another observation is that remnants of earlier races, pre-Mundaric (Austric) and pre-Dravidian, might have formed the nucleus of the Bhil (Koppers and Jungblut, 1976: 4).

These witch songs were collected a long time ago by renowned scholars. Unfortunately, they have so far failed to attract the attention they deserve. They, however, present authentic information required to unravel the mystery of "witchcraft" and its fallout in "witch-hunting," so common among the indigenous people of India. My present preoccupation is to try to understand the meaning of the two "witch songs" and discover what is hidden under the symbolic expressions. The songs are:

1. *I have cut the plantain grove*
   I have taken off my clothes
   I have learnt from my mother-in-law
   How to eat my husband
   On the hills the wind blows
   I have cut the thatching grass
   Weary of eating rice.

   (Archer, 1974: 294, Song No. 437)

2. *The old hag dug a ditch,*
   The Kazaliyo filled it again.
   The old hag filled two pitchers,

The kazaliyo kicked them over.
The old hag climbed on the riverbank
The kazaliyo dragged and threw her,
He pressed her right in the reeds.

(Koppers and Jungblut, 1976: 233. It was collected between 1938 and 1939.)

The first song is a Santhal song, a sort of a soliloquy. The witch is expressing her desire to "eat" her husband as she is weary of eating rice. With this end in mind, she has sharpened her weapon by cutting a plantain grove. For the ritual killing of her husband, she has taken off her clothes. She admits that her husband's mother has taught her the art of "eating" her husband. By cutting thatching grass, she has broken a taboo and declared her animosity against men. Now she is ready to fulfill her desire.

The song delineates the typical image of a witch—naked, hungry, armed with a weapon and the art of killing a human being. This time she is going to eat her own husband. Witches prefer to eat their nearest and dearest ones, including daughters and sons. The time is obviously midnight and the place could be the sacred grove of the village, the crossroad or the village boundary. It is understood that she is not alone, her personal *bonga* or spirit is with her. She will "cut open" her husband's chest and "take out" either all the vital organs, such as, the heart, lungs and liver, or any one of them. She will then go back to where she came from and cook her favorite food and eat it happily. She might share the food with her mother-in-law and her personal *bonga*. The next morning, her husband will fall sick. People will suspect the hand of a witch behind his sickness. The witch-finder (*janguru*) will be summoned. He will identify the woman after going through an elaborate process of witch-finding. Actually those who call him already suspect who the possible offender could be but they do not disclose it before the *janguru*, who simply confirms their guess. The villagers will then ask the woman to confess her guilt and use her witchcraft to cure her husband. She will invariably deny the accusation and protest strongly, but nobody will pay heed to her pleading. She will cry, beg for her life, but will get no sympathy. She might be given a chance to avoid the wrath of the villagers by curing her husband. She might try to ease her husband's pain. If she is lucky, she is saved for the time being. If not, she might be forced to drink human or animal excreta. If she still does not confess her "guilt," people will beat her ruthlessly. Neither her own sons nor the village women will come to her rescue. She might be tried by ordeal.

Trial by ordeal is a severe test of endurance, which the woman accused of being a witch, is forced to pass to prove her innocence. Among the Bhil, the suspect is generally hung up by the arms or by the heels to the branch of a tree and rocked this way and that. While she is hanging, if the branch breaks or if she sustains some injuries like dislocation of the arms or breaking of a leg, she is not considered a witch, but a normal mortal being. If, however, she escapes unscathed, the suspect is a witch (Koppers and Jungblut, 1976: 232). Sometimes she is seized and red pepper is stuffed into her eyes; if this process does not produce tears, the unfortunate creature is condemned; sometimes she is flogged with the branch of *Nux vomica*, or the roots of the Palm Christi, or castor oil plant; and if these make her call out, she is deemed a sorceress, for they alone can inflict pain upon such a being. On other occasions, the witch is tied and put in a bag and thrown into a pool, where sinking is the only proof of her innocence. If her struggles keep her afloat, she is inevitably condemned and punished, either by being obliged to drink water used by leather-dressers, which is a degradation from caste, or by having her nose cut, or being put to death (Malcolm 1880 as quoted by Verma, 1978: 87).

The village which once rejoiced over her arrival as a new bride, the family which in recognition of her worth paid the bride price (*gonong*) to her father, will now express a combination of anger and fear. Nobody will shed a tear for her. There is a strong possibility that she will slowly die the death of a witch, a socially sanctioned punishment for an unpardonable crime. There will be no sense of sin in the minds of the villagers over the gruesome murder.

Witches are accused of "eating" their closest male relatives. Exceptions are negligible. Archer describes only one of such cases out of a total of 15 instances that he documented (Archer, 1974: 294–303). Obviously "eating" is not be taken literally because the "victims" survive after their "vital organs" are taken out. A young witch must qualify for full membership by "eating" someone from her own family. Why does she not "eat" someone from another family (lineage) or another village? According to the song, the mother-in-law teaches her son's wife to "eat" her own son. Taken literally, it sounds strange. So what does the expression "eating" signify? What actually does the mother-in-law teach her daughter-in-law? What knowledge does she pass on to the next generation? Maybe the second song can provide the answers.

The second song is actually a marriage song, however the central character is not the bridegroom, but the witch-finder! The bride is called an "old hag," a common name for witches. The song describes the actions of the witch and the counteractions of the witch-finder, until the witch-finder suffocates the witch to death by forcibly submerging her in the river. The witch kept digging a ditch, and each time the witch-finder filled it up. She filled two pitchers with water, the witch-finder kicked and broke them. Thus, she was not allowed to either dig the earth or store the water. What could be the possible purpose of doing this? The action of digging and watering earth immediately evokes images of agriculture. Was her attempt to cultivate land being opposed? It seems to be so. The conflict finally resulted in her defeat. We found her trying to escape but she was finally caught and put to death. The witch stands for women's right to knowledge and access to cultivation and the witch-finder represents the male dominance over land, knowledge, and agriculture in the community.

Did she commit any crime by digging the earth and storing water for its irrigation? If so, how does that harm the interests of the men? These questions bring us close to only one answer. She, like other women in general, is not supposed to cultivate land anymore. She does not have the right to access land and water, because men have usurped that right. Now there is a new division of labor. Her crime was her opposition to the new order. If this is the message of the song then it is not difficult to understand the purpose of singing this song during the marriage ceremony. Obviously, the purpose is to caution the young bride against nurturing any thoughts about claiming her husband's property. In the rural/indigenous setting, the most important form of property is land and water sources. The song makes sense only if it is seen in this context. It is, in fact, a superb example of the ritual humiliation of women, expressing the male ideology of the indigenous society. The songs belong to two indigenous societies, the Santhal of Jharkhand and the Bhil of Jhabua. Both are well-known for their long tradition of witch-hunting. Anthropologists, missionaries and administrators have collected numerous instances of horrible killing of women on the accusation of practicing witchcraft.

Gender Struggle

The continuing practice of accusation of witchcraft and brutal assault on women in the Ho and Munda society shows, not only men's authority, but also women's considerable access to forces of production and ritual knowledge. If the sex ratio is an indicator of the state of gender relations then the Ho, who register 1,026 females per 1,000 males (Singh, 1994: 404), and the Munda, who register 1,004 females per 1,000 males (ibid.: 843), not only show a relative equality between women and men, but more importantly, an ongoing acute gender struggle in their respective societies.

Both the songs deal with violence. In one, the witch appears to be on a killing spree, and in the other, the witch-finder kills the witch, the "old hag." In the former, the witch is about to kill her husband to "eat" him, and in the latter, the witch-finder kills the witch for a supposed crime. The commonality in both the songs is that both condemn womanhood by depicting women as offenders, antisocials and killers. Ultimately both the witches are killed. In the first song, her death is implied, and in the second, she is killed explicitly. Their killings are backed by social sanctions, by the society where only the male members make decisions and execute them. What sort of a society permits, nay, encourages such killings? Both the songs give us a picture of that society.

The second song associates the "the old hag," the *dakan*, with agriculture. She is killed in a river. Throwing the accused witch into the river in favor of a "purge" trial is also common among the Santhal (Gupta, 1960: 307–19). The alluvial soil of the riverbank is considered to be the original seat of the crops. A woman is normally associated with earth in both *Tantra* and *Samkhya*, as well as in all the forms of indigenous belief systems in India. Her association with water is rather unusual. But if we look at the association from the point of view of agriculture, then it appears to be not only logical but also necessary. And the woman as the inventor of agriculture is now an unquestioned fact. Therefore, the cause of her killing is undoubtedly her association with agriculture. What is being challenged is her control over agriculture. This power of control further enhances her economic and social supremacy. Her control over agriculture gave birth to a social system, embedded in "mother's rights" (Chattopadhyaya, 1959: 258), in the past. Mother's right has been defined as

a form of social organization in which the rights of a person in relation to the other members of his (or her) community and to the community as whole are determined by relationship traced through the mother. In this condition, the duties which a person owes to society, the privileges which he (or she) enjoys and the restrictions to which he (or she) is subject are regulated, and their scope is determined, by the relations in which the person stands to his mother's relatives and his (or her) mother's social group (ibid.: 238).

The earliest form of social organization was centered around the mother. Only the mother could reproduce and nurse the baby. The father was less important in the child's upbringing. The invention of agriculture by women and their knowledge of flora and fauna, their food value and medicinal attributes further strengthened the economic basis of the mother's right. We may therefore, conclude that the song is a condemnation of the economic base of such a social system.

If the second song focuses on the economics of witch killing, the first song refers to the meaning system that supports it. The first song has a clear message of disapproval of the power of the witch. Men feel threatened by such power—the power with which the witch can make or mar the future of men, particularly, in control over land, agriculture and knowledge, in general. It is the "supernatural power" of the witch that men are so scared of. It is considered a mysterious power generated by a secret knowledge. The knowledge is handed down from generation to generation. In this case, the mother-in-law passes it on to the daughter-in-law.[1] The power of the woman is the power of the female principle, the ideology of the social system called the mother's right. Therefore, in the ultimate analysis, the early agricultural economy is the material basis of the female principle (ibid.: 258).

Both the Santhal and the Bhil believe that every woman is a potential witch. Witchcraft is not an individual but a group activity. The Bhils say, *"Kansli matar dakan ane pagri matar sor!"* Every bodice (women's undergarment) is a witch, and every turban is a thief (Koppers and Jungblut, 1976: 231). The Santhal folktale on the origin of witchcraft describes how all the women of the community cheated *Singbonga* and learnt the knowledge of witchcraft from him when he was about to pass it on to their husbands (Bodding 1948: 162–64). Such a belief indicates that, in the past, this knowledge belonged to women in general, and not just to a few individuals as it is today.

What is this knowledge? It is women's power to heal as well as harm. This secret knowledge, not known to men, which gave superiority to women, had to be countered to subordinate women both economically and socially, and establish male domination in society. Men built a parallel system of belief, a different ideology of life, expressed in different ideas and institutions. They overpowered women and implemented taboos to keep them from regaining their lost status and power in society. Thus, both the Santhal and Bhil systems of belief are predominantly masculine. The Supreme Being, the creator of the universe, is presented as a male in the origin myth. Men also hold the institution of priesthood. The assistants of the priests are all men. Women are not allowed to participate in religious rites.

> The belief in the dangerous potentiality of women to seduce evil spirits and wreck vengeance on their enemies is reflected in the fact that participation in all sacrifices is a taboo for women. They are barred from eating the flesh of animals sacrificed to the *Abge Bonga* (family spirit) and *Marangburu* (Supreme Being). A married woman is also barred from entering father's *bhitar* (usually the central room of the house where the ancestral spirits reside) (Troisi, 1979: 221).

Among the Santhal, the women are kept away from the sacred grove (*jaher than*), and are not allowed to be present when sacrifices are offered. Women are also barred from climbing trees of the *jaher than*. Women are regarded,

> as imbued with strange mysterious powers. She cannot for example climb a roof because the nearness of her sex may pollute the *bongas* (spirits). In copulation, the vagina is sometimes described as 'like a *bonga*.' The whole process of attraction—the power of girls to inflame passion, to subdue boys to their will—makes them seem from time to time a source of danger (Archer, 1974: 292).

The taboo on the sexual act before or during auspicious occasions such as rituals, divination or hunting expeditions by men in general and the priests and his associates, in particular, is indicative of two notions. One, woman's proximity on such occasions is considered polluting and two, she wields a mysterious power on men through the sexual act and thereby curb their power. These taboos against women, if seen in association with the economic taboos, such as touching the plough and tilling land or preparation of a straw basket and storing of food grain in it, reveal the true motive behind their operation. It is obviously an attempt to replace a social system where women enjoyed equal if not higher economic power and social status by another, which hastens the process of the defeat of the female sex and establishes patriarchy.

The song further reveals that the immediate area of conflict is the family or the lineage. The witch is accused of "eating" her husband. Eating symbolizes, as understood and depicted by men, the "destructive" power of women. But why do they destroy the nearest ones? Is it an unprovoked act on their part? Men try to establish the idea that women are by nature destructive. Thus, the song says that the witch is craving for the flesh of her husband simply because she is weary of "eating rice." This answer actually hides the fact that the so-called destructive act of the witch is neither unprovoked nor a natural inclination to kill men. It hides the fact that she is engaged in a gory struggle for her survival as a woman. She is fighting against all the taboos imposed on her, and is trying to retain her power and status, and protect the remnants of the social system that sanctions them.

Her cutting of the "thatching grass" is suggestive. Here she challenges the taboo on her right to thatch the house and thereby to own it—a home of her own. But the song depicts her as an offender who is about to break the taboo. In the new male-dominated society, the woman is not supposed to own a home. She is born in her father's home, after marriage lives in her husband's home, and after her husband's death, lives till death in her son's home. In the previous form of society, however, it was a different story altogether. The woman used to live in her mother's home, which eventually becomes her own home too. Her "man" used to pay short visits to her home, but permanently lived in his mother's home. The transition from the one system to the other was neither voluntary nor peaceful.

The song also condemns the witch's nakedness. The witch declares that she has "taken off her clothes." Among the Santhal, the normal practice is that the wife does not undress even during the act of copulation. What can be the possible reason behind it? Is it the desire of the man that the woman should be ashamed of her nakedness—her own body? Is this why the witch is depicted as a naked woman? It is believed that she, in her witch form, dances naked, moves around naked, copulates with her *bonga* naked, cooks and eats the vital organ of her victim naked. Therefore, she is antisocial. This view denotes the conflict between the "system" and the "anti-system." Patriarchy in the Santhal society is almost an established fact today. If this is the system, then the anti-system is fighting the last battle. It has gone underground to a large extent. The world of the witch is the netherworld, a secret realm, the domain of the antisocials. The song in its totality shows, though unintentionally, that the woman's spirituality is under male attack. Whatever she is not allowed to do or perform in her real life, is executed in the realm of her witch self. The belief related to the witches tells us how the witch breaks all the taboos. She has her own set of spirits who are labeled "evil spirits" by men. The men's benevolent spirits are used by the witch-finder to catch the witches. The woman is barred from the sacred grove despite the fact that its ruling spirit is a female spirit, the *jaher era* (the woman of the sacred grove). The witch goes there and cooks the organ of her victim. This indicates that she performs rites of her own belief there.

The two witch songs form a part of a myth, the myth of witchcraft. All the myths, folktales, and songs related to the witches are created to perform a social function. One should, however, be careful of not falling into the trap of considering them to be history. "Myths may be part of culture, history in providing justification for a present and perhaps permanent reality by giving an invented "historical" explanation of how this reality was created" (Bamberger, 1974: 267). These "witch songs" and other tales that we have mentioned themselves do not give us a bit of history. On the contrary, they actually create a false reality. But at the same time, they cannot escape the historical truth hidden behind it—a fierce and bloody struggle that has been going on between the female and male genders in these societies. Witch-killing is an expression of that conflict—the gender struggle. If women comprise the first economic class, then gender struggle is the first form of class struggle (Nathan, 1997).

Both songs represent the ideology of growing patriarchy as opposed to the female principle, the ideology of the "Mother's rights."

> In *Tantra* and *Samkhya*, the female principle or *prakriti* is viewed as the fundamental reality, the cause of the universe. Such an idea could only be the reflection, in the sphere of philosophical abstraction, of the social supremacy of the female (Chattopadhyaya, 1959: 233).

Patriarchy emerged in the process of destruction of the social and economic foundation of the female principle. The causes of the growth and decay of the social system of mother's rights and the emergence of patriarchy on its debris thus hold the key to unravel the mystery that shrouds the phenomenon of witch accusation and witch killings.

We are not, however, sure of the exact form that female supremacy assumed in the dawn of human society. Matriarchy as Bachofen (1967) portrayed may never have existed. However, the last generations of our Indian historians comprising giants like Kosambi (1985) and Chattopadhyaya (1959) believed in the existence of a modified form of matriarchy. Today even that modified version of matriarchy may also be challenged. But several matrifocal and matrilineal societies still exist in our modern world. Numerous myths and folktales testify the existence of such societies where women hold relatively equal, if not superior, status. Therefore, patriarchy is neither natural nor the earliest form of social organization. Men constructed it to rule over women, to subjugate them sexually and physically—to exploit their productive and reproductive labor.

NOTE

1. The Azande believe that witchcraft is a physical trait transmitted by unilineal descent from parent to child (Evans-Pritchard and Gillies, 1976). Mahapatra, however, thinks on the basis of his fieldwork in Mayurbhanj in Orissa that the Santhal do not believe in hereditary transmission. On the contrary, witchcraft is looked upon as an art, a discipline whose essential features are learning of *mantras*, regular and exotic, physical and spiritual practices, austerity, and purity. It is like any other art, which is to be learnt, which may, no doubt, be enhanced by personal factors (Mahapatra, 1986: 101). However, this observation does not reject the possibility of the existence of the practice of hereditary transmission of such knowledge in the remote past.

REFERENCES

Archer, W.G. (1974) *The Hill of Flutes: Life, Love and Poetry in Tribal India: A Portrait of the Santhals*, University of Pittsburgh Press, Pittsburgh PA.
Bachofen, J.J. (1967) *Myth, Religion and Mother: Selected Writings of J.J. Bachofen*, Princeton University Press, Princeton, New Jersey.
Bamberger, Joan (1974) 'The Myth of Matriarchy,' in Michele Zimbalist Rosaldo and Louise Lamphere (eds.), *Woman, Culture and Society*, Stanford University Press, Stanford, pp. 263–80.

Bodding, P.O. (1948 [reprint 1994]) *Traditions and Institutions of the Santhals: Horkoren Mare Haparomko Reak Katha*, Bahumukhi Prakashan, New Delhi. (Rev. L.O. Skrefsrud published the original Santhali text in 1887.)

Bosu Mullick, Samar (1991) 'The Concept of Supreme Being in the Sarna System of Religion,' in Bosu Mullick (ed.) *Cultural Chotanagpur: Unity in Diversity*, Uppal Publishing House, New Delhi.

Chatterji, S.K. (1970) *The Origin and Development of the Bengali Language*, George Allen & Unwin, London. First published in 1926 by Calcutta University Press.

Chattopadhyaya, Debiprasad (1959) *Lokayata: A Study in Ancient Indian Materialism*, People's Publishing House, New Delhi.

Dalton, E.T. (1872 [reprint 1960] *Descriptive Ethnology of Bengal*, Thacker, Spink & Co., Calcutta.

Evans-Pritchard, E.E. and **Eva Gillies (eds.)** (1976) *Witchcraft, Oracles and Magic Among the Azande*, Oxford University Press, London.

Gupta, R.K. (1960) 'Witchcraft Murders in the Duars,' *Man in India*, 40(4):309–19, Ranchi.

Hoffman, J.B. and **Arther van Emelen** (1950) *Encyclopaedia Mundarica*, Vol. I-A, *Asur-Kahani*, pp. 240–50; Vol. II-B, *Baranda-bonga*, pp. 422–33, *Ba-porob*, pp. 383–90; Vol. IV-D, deonra, pp. 1022–39; Vol. X-M&N, *najari* and *najom*, pp. 2913–22, Nage-era, pp. 2906–7; Vol. XIII-S, *Sokha*, pp. 3033–4034, sengelda, pp. 3918–22, *Sing-bonga*, for the origin and myth and other legends, pp. 3981–88. Government Printing Press, Patna.

Kelkar, Govind and **Dev Nathan** (1991) *Gender and Tribe: Women, Land and Forest in Jharkhand*. Kali for Women, New Delhi.

———— (1993) 'Women's Land Rights and Witches,' in Mrinal Miri (ed.), *Tribal Society: Continuity and Change*, Indian Institute of Advanced Study, Shimla, pp. 109–18.

Koppers, W. and **L. Jungblut** (1976) *Bowmen of Mid-India*, Vol. 1, Wien, Austria.

Kosambi, D.D. (1985) *An Introduction to the Study of Indian History*, Sangam Books, Pune.

Mahapatra, Sitakant (1986) *Modernization and Ritual: Identity and Change in Santhal Society*, Oxford University Press, Calcutta.

Malcom, Sir John (1880 [reprint 1923]) *A Memoir of Central India Including Malwa and Adjoining Provinces*, Vol. I, Parbury & Allen, London; Vol. II, Thacker, Spink & Co., Bombay, and W. Thacker & Co., London, 1880.

Nathan, Dev (1997) 'Gender Transitions in Tribes,' in Dev Nathan (ed.), *From Tribe to Caste*, Indian Institute of Advanced Study, Shimla.

Rawson, Philip (1978) *The Art of Tantra*, Thames and Hudson, London.

Risley, H.H. (1891[reprint 1981]) *The Tribes and Castes of Bengal*, Vol. I, Firma Mukhopadhyay, Calcutta.

Roy, S.C. (1912 [reprint 1970]) *The Mundas and Their Country*, Kuntaline Press, Calcutta. (Reprint: Asia Publishing House, Bombay.)

Russel and Hiralal (1969) *The Tribes and Castes of the Central Provinces of India: Vol. 3*, Anthropological Publications, Oosterhont. N.B., Netherlands.

Singh, B. (1977) 'Aspects of Kodaku Culture,' in Ajit K. Danda (ed.), *Tribal Situation in North-East Sarguja*, Anthropological Survey of India, Calcutta.

Singh, B. and A.K. Danda (1986) *Kodaku of Surguja*, Anthropological Survey of India, Calcutta.

Singh, K.S. (1994) *The Scheduled Tribes: Anthropological Survey of India*, Oxford University Press, New Delhi.

Sinha, Surajit (1961) 'State Formation and Rajput Myth in Central India,' *Man in India*, 13(2), Ranchi.

Troisi, J., (1979) *Tribal Religion: Religious Beliefs and Practices Among the Santals*, (South Asia Books) Manohar, New Delhi.

Verma, S.C. (1978) *The Bhill Kills*, Kunj Publishing House, New Delhi.

World Book—Multimedia Encyclopedia (1997) Published by World Book Inc. and IBM Corporation, New York.

II

FOREST MANAGEMENT

VI

Forest Management in Mosuo Matrilineal Society, Yunnan, China

HE ZHONGHUA

To study the Mosuo matrilineal system, the management of the forest, and the influence of the recent policy of devolution of ownership of the hills and forest, my associate He Fengju and I spent 24 days from 13 July to 6 August 1999 carrying out investigations in Luoshui and Zhengbo villages in Yongning plateau of Ninglang county in Yunnan province. Both villages are natural villages where the Mosuo matrilineal system prevails. Their common characteristics are:

- Matrilineal households (including "double-system" families) comprise 90 percent of total households.
- The population is mainly Mosuo, with a small number of people of other ethnicities.
- Mosuo culture occupies the leading position and influences the other ethnic minorities.
- Gender relationships are relatively equal and harmonious, but as far as social status is concerned, men are superior to women.

- Women play a positive role in the protection and utilization of forest resources.
- Because of excessive deforestation in the past which has caused ecological imbalance and calamities, the ecological consciousness of the people has been greatly enhanced.

However, there are also differences between the two communities:

- Luoshui village is becoming rich by taking advantage of tourism, whereas Zhengbo village, as a result of following its traditional way of life, is not.
- Luoshui village is situated in the natural protective zone and the forest is well protected, whereas Zhengbo village continues to make a living from deforestation, and so its ecology and environment remain poor.
- Luoshui village does not collect much firewood, whereas Zhengbo village practices very severe cutting of trees for firewood.
- Tourism in Luoshui village has brought about rapid economic development and less division of labor between women and men. However, in people's relationships in general, and in the relationship between women and men in particular, some negative influences have been felt. In contrast, Zhengbo village continues to observe its traditional way of life, thus the villagers enjoy their traditional habits and customs as usual, and the relationship between women and men is relatively balanced.

In view of the above circumstances, we forward our hypotheses on the relationship between (a) economic development and forestry development, (b) devolution of forest rights and forestry development, and (c) the matrilineal system and forestry. These hypotheses are: With economic development and an increase of income, people will ask for less of the forest, thus promoting the ecological service function of the forest more effectively. On the other hand, a backward economy will make people ask for more of the forest, which will affect the protection of the forest, thus weakening its ecological service function. The implementation of the policy of devolution of rights is positively related to the protection of the forest and the life of the villagers. The matrilineal system and religion play a very positive role in the effective protection of the forest in the Mosuo region.

Historical Perspective on Forest Policy

Ninglang county is situated in Yunnan's northeast corner bordering Sichuan province. At elevations of between 1,350 m and 4,510 m, its three-dimensional topography and mild climate provide an advantageous environment for the growth of diverse forest and plants; hence the county is known for its rich forest resources (*County Annals of the Ninglang Yi Autonomous County*, 1993). In 1989, forest cover was 53.4 percent, totaling 3,980,000 *mu* (1 *mu* = 0.07 ha). The average forestland per capita was 19.75 *mu*, and the average live timber per capita was 152 m³, the highest rate not only in the province, but in all of China.

Forestry is the principal industry of Ninglang county. Between 1949 and 1993, cut forests totaled 1 million *mu* producing 3,619,000 m³ of timber including marketed timber, firewood and timber for household use, making up over 70 percent of total tax revenues of the whole county (*Forest Annals of the Ninglang Yi Autonomous County*, 1995). Such excessive cutting of the forest has had serious consequences. Forestland is decreasing and this is coupled with disastrous soil erosion and frequent natural calamities. The government has only begun to pay attention to afforestation in the last decade. Rampant deforestation still continues unchecked despite official decrees prohibiting it, and lumbering remains the major source of income for both the government and peasants in the mountain areas. Recently, the government initiated a total ban on lumbering in the upper valley of the Yangtse river, yet there is no other industry to replace it.

Before the founding of the People's Republic, the hills and forest in Ninglang belonged to feudal lords or *tusi*. In 1956, democratic reforms began to transfer the ownership of the hills and forest to the state. In 1961, the county government divided the forest into two categories: state-owned and collective-owned, either by the administrative district or the production brigade. The former made up 21.4 percent, and the latter 78.6 percent of the total forested land. During the next redistribution of forestland in 1962–63, large patches of natural forest went to the state while production brigades owned no forest as collectives. However, they enjoyed the right to care for patches of state-owned forest zones, including forests close to reservoirs or headwaters and fruit tree plantations planted by commune members around their houses or in their household garden plots. This redistribution resulted in the government owning 69.8 percent, and the collectives 20.2 percent of the total forested land.

During the Cultural Revolution from 1965 to 1979, all collective-owned forests were transferred to state ownership. In 1980, the principles for the implementation of the ownership of the hills and the forest were changed again, returning to the rough pattern of ownership created in 1961. Namely, large patches of natural forest were still state-owned, and the collective-owned forest taken by the state from 1963 to 1966 was returned to village production brigades. Production brigades close to the forest who had only taken care of the forest, but not owned it, were now given nearby plots of forest for collective ownership. When available, some pieces of barren hills or fields were given to individual commune members for private use, with each person receiving 1 or 2 *mu*. This readjustment of forest ownership meant that state-owned forests accounted for 44.8 percent of the total against 55.2 percent owned by the collectives in 1980.

In 1982, a forestry policy was implemented which stabilized the ownership of the hills and forest, finalized the allocation of hills and forests for private use, and defined duties and responsibilities in forestry production. Meanwhile, a policy was also enforced to fix the total area under cultivation and transform cultivated areas back into original forestland or pasture.

In 1985, Ninglang delimited responsibility hills and private-use hills. Forests in private-use hills belonged to the collective, but the mineral resources belonged to the state. The right of the management of the forest belonged to the peasants, who could plant and harvest trees or grass at their convenience. The policies were not to change over a long period of time, and the right of ownership could be inherited by descendants. The rights to ownership of the hills and the forest, lumbering and selling of responsibility hills belonged to the collective. Private-use hills and responsibility hills now made up 30.40 percent and 69.60 percent of the collective-owned forest, respectively.

In 1992, certificates were issued for the state-owned hills and forests, and quotas for cutting of the forest were specified. Decrees were also issued that as a principle, protected forest, headwater forest, scenic forest, forest in the natural protective zone, and ancient or rare and precious trees should not be cut. Afforestation and lumbering were to be conducted in an orderly and planned manner.

In short, from 1959 to 1993, many adjustments of ownership were made, which resulted in the allocation of part of the forests to collective ownership. Although the question of ownership has now been finalized, management of the hills and forests has been initiated, and attention directed to afforestation, deforestation remains unchecked. While deforestation has resulted in frequent mud-rock flows and floods, the state, collective, and privately owned timber companies continue to go about their business as usual.

According to Guo Huaizhong and Yang Shengming, head and secretary respectively of the Yongning District Communist Party, the Mosuo in the county mostly live in valley floodplains. Therefore, forestry was not of great importance to them in economic terms, although people did use timber to build houses and collected firewood from the forests. As a result, according to the policy adjustment in 1982, and based on the availability of nearby forests, most Mosuo villages were not given forest hills for private use. Instead, they were symbolically given a certain part of the responsibility hills.

The relationship between the Mosuo communities and logging in Yongning today is as follows. Timber for use by local villagers (mainly for housing construction) can be cut only from the community forest after obtaining approval of the village committee and the Forest Management Station of the township. State-owned forest enterprises and the forestry bureau are allowed to log in the state-owned forest, and they should comply with quotas approved by the provincial and county governments. Villagers in or near forested areas can log and earn some income. Upland Yi people who live in forested areas close to the state-owned forest have more opportunities to work as loggers. In addition, Yi people have traditionally been nomads engaged in slash-and-burn cultivation, and they have their own forested land. This accounts for the fact that many of them are engaged in logging. However, Mosuo people have traditionally had various taboos against the cutting of trees. Their matrilineal social system also made it impossible for them to have as many extended families, and they therefore did not require as many logs for building new houses. So, generally speaking, the Mosuo people do not make much money by working as loggers, and cannot benefit much from the timber trade.

Sociopolitical Conditions of Mosuo in Yongning District

Yongning district is situated in the northeast of Ninglang county, 93 km away from the county seat. It borders on Sichuan's Muli

Tibetan autonomous county in the north and Sichuan's Yanyuan county in the east. The total area is 641 km², with an average elevation of 2,644 m, and total population of 17,081 at the end of 1998. Yongning has six administrative villages and 64 natural villages. Within its territory are numerous mountains, and the Yongning plateau, a large flatland with an elevation of over 2,600 m. Lugu lake, near which the villages in this study are situated, is one of the few deep lakes on the plateau still free of pollution. It was listed as a "Grade II National Natural Protection Zone" in 1989. In Yongning plateau and around the lake live Mosuo who still practice a matrilineal system, whose population, according to statistics at the end of 1995, was about 6,160.

Before 1949, Yongning was a society of feudal lords, whose highest ruler was the *A's tusi*. Their society was classified into three ranks: the *sipi,* the *zeka,* and the *e*. The *sipi* were the nobility, the *zeka* were the common people, and the *e* were the slaves. The *tusi* practiced the system of primogeniture. During the rule of the Chinese Republic, the *bao-jia* system was enforced, but had little effect. The *tusi* ruled by its conventional laws, and all the land, forests, and other natural resources were owned by the *tusi* and *sipi*. In 1949, the *tusi* system was abolished, and the administrative organizations became the current *xiang* (administrative district), the administrative village, and the local community.

The economy of the Mosuo in Yongning is mainly farming, with rice, millet, maize, wheat, highland barley, potato, buckwheat, oats, and various species of beans as food crops. Animal husbandry and horse transport provide some secondary income. Domestic animals include cows, horses, goats, and pigs which are for home consumption, not for sale. The Mosuo cottage industry includes linen making, production of woolen goods, breweries, and oil pressing. Handicraft production is not yet separate from farming, and so their economy remains in a state of self-sufficiency. Nowadays, people wear ready-made clothes bought in the market, and only very few aged people know their traditional craft of weaving.

For a long time, the Mosuo in Yongning district have practiced a family and marriage system completely different from the traditional Chinese patrilineal system; namely, the Mosuo's matrilineal system and "visiting marriage" known colloquially as *axia*. In fact, no marriage system exists at all and Mosuo adults do not get married at any time in their lives. Once in love, they work and live in their own homes in the daytime, and at night the man goes and stays with the woman as an overnight visitor, leaving early the next morning. After cohabitation, if they have any children, these children will be brought up by their mother as her own family members. The man, for his part, undertakes to bring up and educate not his biological children, but the children of his sisters in his role as their "uncle-in-law." As for his own children, he may occasionally buy them some clothes or give them pocket money if he can afford to, but this it is not obligatory. His children may visit him only if he is ill, and have no other regular visits with him. Therefore, their affection for their uncle-in-law is deeper than for their father.

The Mosuo value their matrilineal family structure because it advocates equality between women and men. They respect the old and love the young, encourage unity among the people, and amicably put up with each other. They observe the tradition of *axia* because this kind of "marriage" offers more equality to both sexes, and more freedom to human nature, free from the fetters of mercenary marriage or family status. Because of this, the matrilineal system and its *axia* have been able to eliminate political, economic, and cultural interference from outside, and flourish through the ages. In modern times, however, owing to its historical background, geographic location, and outside cultural impact, the matrilineal system is undergoing various large and small changes in different communities.

Generally speaking, in political and economic centers where outside cultural influences are stronger, and in the places where *tusi* culture still exerts an influence, the matrilineal system is not as strong as it is in more remote border districts, where it is still very pronounced. In Yongning's Mosuo society, in addition to the matrilineal system, there is a patrilineal system and, in some families there is a coexistence of both matrilineal and patrilineal systems, which ethnologists would call a "double-system" family. A patrilineal system exists in the *tusi* families and in the nobility, as well as in families which are short of male labor or in the families of women slaves or maids whose marriages were arranged by the *tusi*. During the Cultural Revolution, a large number of people were forced to get formally married. Families formed in this way also practice the patrilineal system. The double-system family also exists where the wife and husband prefer to get formally married. It also happens when a member of a patrilineal family wants to practice *axia*.

The cultural feature of the matrilineal system can be seen prominently in the relative equality between the two sexes, in the observation of natural unions, in the strong worship of motherhood and collectivism, and in the worship of a mixture of deists and Lamaist deities.

Luoshui Village

Luoshui village is located by Lugu lake to the southeast of the Yongning plateau, 20 km away from the seat of the district government. It is part of the larger Luoshui administrative village. The village is divided into two parts, the upper village and the lower village. The former sits next to the mountain slope, and its inhabitants are mostly Pumi. The latter is located near the lake, and its inhabitants are mostly Mosuo.

According to late 1998 statistics, there were 76 households comprising 485 people in both the upper and lower villages, of which the Mosuo constituted 33 households and the Pumi 21. The rest were the Hans and the Bais. During my investigation in the lower village in 1996, I found that of the 33 Mosuo households, 31 (94 percent) were matrilineal families. Pumi in the upper village commonly followed the *axia* practices of the lower village Mosuo. Our last investigation in 1999 found that the Pumi households in the upper village had increased to 25, and that 16 (64 percent) of them were matrilineal. In fact, except for maintaining their own language and a slightly different calendar year, Pumi follow all the customs and habits of the Mosuo.

The traditional productive activity of Luoshui village is farming, which is also the major source of trade. However, with poor soil, low temperatures and inferior irrigation, cultivable crops are limited to maize, potato, buckwheat, and beans. In the old days, the home production of food could not meet the grain consumption needs of the village, and so people had to exchange dry fish for grains in Yongning town each year. In the early 1990s, the villagers had to eat subsidized grain from the government. The average per capita income being only around 200 yuan, most of the people would go without grain after the spring festival. As a result, Yongning was known far and wide as a poor area. However, the past few years have seen much improvement. The villagers have had very productive farming, and there has been no drought. Each person now has approximately 1,000 kg of grain, and food is no longer a problem.

Fishery is the second most important productive activity after farming. Lugu lake was once rich in fish. The old people would tell you that once upon a time, when there were a lot of fish in the lake and in the rivers, you could easily catch a basketful of them with a dustbin. These were mostly fine-scale fish. During the rule of the *tusi*, only a small tax was levied (30 to 40 pairs of fish per household), and you could exchange fish for rice. For the past 30 years, however, because of the introduction of new species of fish mixed with some harmful species, the construction of a power station at the outlet of the lake, and predatory fishing by certain people, fish resources are becoming very scarce. For the past few years, tourism has become the main source of income, and the formerly rich fishery amounts to almost nothing.

Animal husbandry holds some weight in the village economy as well. People raise cattle, horses, sheep, pigs, and chickens. Cattle are draught animals, while sheep, pigs, and chickens are raised for subsistence meat. Horses, which used to be raised as pack animals, are now a source of money in the tourist trade, where they are put into service for horse rides.

Since 1991, taking advantage of the natural beauty of Lugu lake, the Luoshui villagers have opened a lot of family hotels complete with boating, horse rental, song and dance evening parties, and other tourist entertainment. By 1996, tourism had replaced farming as the leading industry of the village and was the main source of income for individual households. In 1996, annual per capita income was 2,000 yuan, and the collective village income from boating, renting horses, and other tourist activities alone came to 100,000 yuan. Luoshui is now known far and wide as a relatively wealthy tourist village and is, in fact, today one of the 10 richest villages in Lijiang prefecture.

With the development of tourism, drastic changes have taken place in transportation, communications, housing, and other local infrastructure. The way of life and the mode of thinking have also undergone important change. Now the former dirt highway to the district seat is asphalt. Once impassable, the road around the lake can today reach Yanyuan county town in the dry season. Thirty-two out of 33 households have built new wooden houses. Telephones, refrigerators, washing machines and television sets, and small appliances are increasingly popular. Each family has a flush toilet, and some households even have water heated with solar energy equipment. Accordingly, the mode of thinking is changing notably. People are becoming aware of doing business and have a rough idea about interpersonal competition. However, they still retain their matrilineal system, *axia*, funeral customs, and sacrificial rites. In fact, they have even further consolidated these practices to attract tourists.

Gender Division of Labor

In the division of labor between women and men in Luoshui village, the villagers follow their traditional mode, namely "men work in the fields, and women do the housework." However, this division has developed into a new phase, in which men participate in public affairs and business, for instance, soliciting customers and loans, while the women handle the reception of tourists, besides doing housework. Moreover, with the development of tourism-related businesses, men are beginning to play the leading role in important decisions.

In our investigation of some of the larger family hotels, though the tradition is still observed that the whole family makes the decisions on important matters, these decisions are forwarded and executed by the men. Take, for example, the Mosuo Garden Hotel, run by Shize and his sister Zelacuo. Shize decides on all the important matters including handling loans and expanding the business (which now encompasses a hotel in town run by their brothers). Zelacuo is busy playing the role of receptionist for tourists. Or, take the Mosuo Scenic Garden Hotel also run by a man and his sister. In this instance, the man is the head of the village. Besides handling village affairs, he raises important general decisions about the hotel for approval by the whole family. As to strictly business decisions, the brother, whose name is Ruci, makes these because it is his designated job to do so. His sister Danshizhima takes care of the food and lodging for guests and does housework. Likewise, Family by the Lake Hotel is run by a man from outside who is married to a girl of the village. A clever young man, who is good at business, has taken charge of the business. When he arrived, he immediately won the confidence of the family and seized the leadership of their hotel business. This is something that often happens to strangers who join Mosuo as family members. In Yongning, more such instances can be found.

All these examples show that as a result of the expansion of ties between Mosuo households and the larger society, and the close connection between household production and social production, men who have a traditional role in the social arena are offered an opportunity to give full play to their superiority, thus expanding their power, whereas the women, whose traditional work is confined to household labor, find it difficult to rise in the same arena. The presence of men from other ethnic nationalities, brought up in the system of patrilineal systems and possessing strong business abilities and economic skills, could naturally and imperceptibly remold Mosuo traditional culture, and this constitutes a crisis for Mosuo culture.

On the other hand, service scope has widened with the development of tourism, and this offers opportunities for both genders to engage in the tourist trade. As a result, the division of labor has become narrower and the sharing of labor power between women and men more equal. Earlier, in the gender division of labor, men did the heavy work, and women did relatively lighter but time-consuming work. Nowadays, in the tourist service, division of work does not exist in many areas of labor, for instance, in boating, in leading a riding horse for a tourist, or in singing and dancing performances. Women have to do housework all the same, which leaves them with less spare time. Yet, these days men are beginning to collect pine needles in the hills for fuel. Nevertheless, when it comes to collecting firewood, men use horses or tractors to haul the wood back home, while women carry it themselves. A time study made of the daily work of women and men in the Waru family of three working people shows that there is almost no gender difference (see Table 6.1).

Table 6.1
Time Study of the Waru Family Members

Time	Suji (Mother, 62)	Luruo (Son, 39)	Nuozhu (Daughter, 32)
0600–0700	Boiling water, burning joss sticks	Walking the horse	Carrying water, cooking
0700–0800	Breakfast	Breakfast	Breakfast
0800–1300	Feeding cattle, cooking pig feed, feeding pigs, washing WC	Boating or renting horses	Farming or cutting firewood
1300–1400	Lunch	Lunch	Lunch
1400–1800	Cutting pig grass, feeding pig, burning joss sticks, driving cattle home, cooking supper	Boating or renting horses	Farming or cutting firewood or grass
1800–1900	Supper	Supper	Supper
1900–2200/2300	Meditation, lighting Buddhist lamps, kowtowing	Visiting, singing and dancing at a party or resting	Visiting, singing and dancing at a party or taking care of baby
2200	Sleep	Sleep	Sleep (at 2300)

In the realm of religion, the inequality between women and men appears more distinct. The Mosuo have their own religion, *daba*, which is full of masculine color. The monks, who are considered to be the envoys of God, succeed from father to son and uncle to nephew. Even the shamans are males. Likewise, in Mosuo Lamaism, women are forbidden from participation in a lot of activities. For instance, in important sacrificial rites or funerals, women are not allowed to enter the scripture hall where the Lama chants the scripture. Women are not allowed to wash the corpse of the dead, and during cremation, they are allowed only to stand downhill watching from afar. Even in daily life, the saying goes: "Men each have a piece of holy hair, but women do not. They have nothing in the wombs of their mother, and so they are doomed to be born women." Men also chair all the sacrificial ceremonies and work as assistants to the scripture-chanting Lamas, receptionists, and sacrifice carriers. Women can only kowtow, make sacrificial articles, cut and send packs of green pine needles to sacrificial places, or help the men cook. In all sacrificial ceremonies, women do the odd jobs, which contrasts sharply with their leading role at home.

Men have also grasped social power. During the rule of the *tusi*, the power was in the hands of men, and the sons succeeded their fathers. After the democratic reform, people began to seize political power. However, men were the main political leaders. Until now, the seat of a woman chair has remained vacant for three years in the Yongning district government. In the former Village People's Committee of Luoshui village, the ratio of men and women committee members was 6:0, and there was no seat for a woman committee director. Last April, the committee began to offer a woman director's seat, but the ratio of men to women was still 8:1. Women are always considered to be of poor ability or less educated, incapable of participating in political affairs. In reality, some of the women are no less qualified, and some of the men are no more educated, yet the seats of village heads or officials are always for those said to be abler. No opportunities for women! Our investigations were persuasive on this point.

Changes in Use and Ownership of Forests

The ownership of the forest in Luoshui underwent changes during two different epochs. During the time of *tusi* (before 1956), the forest and the land belonged to the *tusi*, who managed them according to their conventional rules and regulations, which mainly were:

- From May to August each year, cutting trees or bamboo in the hills was not allowed. It was believed that cutting at this time would cause hail and damage the crops.
- Black Yi sheep owners had to pay a grassland tax, generally one or two sheep for a flock of sheep.
- Deforestation was absolutely prohibited in holy hills, water resource forests, scenic forests, graveyard hills, and bone-ash hills.
- In Yongning plateau, people who cut any trees in the hills had to give a meter or more of wood to the *tusi*.
- Religion advised the people to do good things, plant trees, and protect the forest. Living trees were not to be cut, only dead ones.
- Severe punishments awaited those who dared to deforest the hills.
- Anyone who dared to violate the rules and regulations would be punished by having to butcher an ox or a pig. If offenders did not behave well, they would be punished severely, including lashes on the hips.

Since the democratic reform in 1956, the ownership of the forest has undergone many drastic changes: State ownership; state ownership + collective ownership; state ownership + collective ownership + private-use hills; state ownership. After the democratic reform, all the hills and forests were nationalized. In 1962, there was collective-owned forest, and in 1982, there were peasant-owned private-use hills. In 1983, The Lugu lake natural protection zone was founded, and both the collective-owned forests and private use hills were taken back to state ownership. In early 1988, the Lugu lake natural protection zone authorities enforced the policy of closing hillsides to facilitate afforestation. To solve the problem of firewood, a plot of land for a firewood hill was given to the villagers in Hongyanzi, about 4 km away from Luoshui village.

Owing to the fact that the matrilineal system does not break up the family; new houses are not needed. However, when tourism began to flourish, from 1994 to 1996, the peasants in the lower village were in a rush to build new houses, which resulted in an acute consumption of timber. In view of the actual need, the county and the district governments allotted a timberwood hill about 70 km away in Labo administrative village as a collective-owned hill for Luoshui village. It was, however, too far away from Luoshui village for effective management. Now it is said that there is no more timber, and the villagers have to buy timber from Sichuan province or elsewhere in the county beyond the natural protection zone.

In the past, villagers' utilization of forest was principally subsistence-based consumption which did not waste much of the forest resources, and overall the forest ecology was not in a bad state. Timber from the forest was used to build boats, houses and farm implements, bamboo served as construction material for many household articles, and fungi and forest plants were harvested for medicine and food. According to the old people, the forest around Lugu lake was so dense that one could hardly get through it, and that it had leopards, river deer, *muntjac*, and tigers. The *tusi* exercised very strict management. No deforestation or hunting of wild animals was permitted. Until the highway was opened in 1972, there was dense forest. After that, 1973 and 1974 saw large-scale timber logging by timber companies, and all the trees near the highway were cut down. For the past decade, the government has closed the hillsides for afforestation, and a gradual recovery of the vegetation can be seen. In addition, forest space is expanding as a result of the returning of grain plots to forestry. Lugu lake has green woods far and near. Natural disasters are becoming less common, and water sources are improving. What is very important is that these changes have created superb conditions for forest ecological tourism. Villagers' income has increased considerably due to tourism, and their personal experiences have promoted their need to protect the environment. They are beginning to demand less of the forest, thus forest ecological protection is encouraged.

So far as we know, the reason why forest ecological conditions were so good in earlier times, apart from strict conventional rules and regulations exercised by the *tusi*, can be largely attributed to religious strictures, which may be summed up as follows:

Holy trees, 1,000-year old trees (the few giant trees on a hill), and water-source trees could not be cut. Anyone who dared to cut those trees would suffer. During the Cultural Revolution, in the name of "doing away with superstition," groups of people were ordered to cut these trees. The woodcutters were heard to speak to the trees, saying "Forgive me. It is not I that want to cut you down. Damn the officials!" After cutting down the trees, the woodcutters would cover the tree stumps, so that God would not see them, thereby hoping to avoid punishment.

Young trees could not be cut. The cutting of young trees was considered to be like killing one's own children. If you cut young trees, you would then be without offspring. If, when felling larger trees, you damaged any young trees, you had to try and make the small trees stand upright again. White trees (white-barked trees) were also not allowed to be cut. Cutting them, it was said, would bring about hailstone storms. If you cut a pig trough from white timber on a hill, you had to color it red with mud before bringing it home. Big trees cut down as firewood must not be cut open so as not to reveal their white color inside. When deliberate violation of these rules coincided with a hailstorm, the violator would be accused of the crimes and asked to butcher pigs or sheep to provide a feast for all the villagers as a punishment.

There were also some religious activities undertaken each year to pray for the safety of both humans and cattle, and to invoke a bumper harvest. Among these were the sacrificial rite *toyila*, conducted in summer and autumn; the monthly walking around the hills, and the offering of sacrifices to the hills on the fifth, 15th, and 25th day of the lunar calendar; and, the daily turning of the scripture barrels. When doing so, people would chant and pray to the gods of the earth for their blessing. These religious activities were objectively helpful to forest ecological protection.

With ownership of the forest now transferred to the state, the management and protection of the forest is conducted in two ways. The first is direct management by the Lugu Lake Natural Protection Zone Committee, whose job it is to protect the forest from fire and close hillsides for afforestation. Four forest guards are hired to watch the hills, each at 150 yuan a month. The other way is the signing of contracts with villagers in the protection zone. So that villages might still profit from the protection of the forest, they are allowed to collect dry and dead trees as firewood, and fallen pine tree needles as cushions for cattle sties as long as they do so in a specified area from the protection zone to Hongyanzi. They are forbidden to cut any trees, living or dead; nor can they cut grass. In Luoshui village, the villagers stipulate that anyone who cuts a big tree be fined 200 yuan, and in the case of a small tree, 50 to 100 yuan. One can only collect firewood on the hillsides of the Red Rock Hill, and then only dry trees and twigs. A peasant household in the upper village which secretly cut down five big trees, when discovered, was fined 1,000 yuan. As an apology, they also offered a bottle of wine and a chicken.

Before tourism rose in 1992, Luoshui village was a location for social forestry financed by the Ford Foundation. By 1994, when tourism had become increasingly prosperous and villagers found that they could gain fast and large profits, the Ford project was no longer attractive to the villagers. The project almost came to a standstill. Later, it was transferred to Fanglang and some other poor communities. At present, a Japanese agricultural tourist village project has chosen Luoshui village as a site for a project which was scheduled to begin before the end of the year 2000.

In terms of the current management of forest resources, since Luoshui village is located in the Lugu lake nature reserve, its forest resources are owned by the state and controlled by the state forestry bureau. The villagers can participate in forest preservation and management by signing contracts with the reserve. When they cut down trees, villagers will usually cut those which are dry, and dry fallen pine needles and firewood can be collected in the forest. However, in the part of the forest extending from the nature reserve to Hongyanzi village of the Luoshui administrative village, collecting of firewood and pine needles is not allowed. This means Luoshui villagers have to travel about 4 km to collect their fuelwood. To cut down trees, they must go to Labo administrative village 70 km away, and the long distance makes it impractical for most people to go and cut timber there.

With the development of tourism, there has been an increasing demand for firewood from the local villagers. However, this has not led to excessive cutting of forest resources in the local area. Instead, villagers deal with the demand by purchasing waste and used building materials from nearby wood storage and timber plants, using electricity and hay straw as fuelwood substitutes, and adopting firewood-saving ovens. They also buy firewood from the Yi people living in the mountains, and timber from the nearby villages which are not within the nature reserve. It is evident these last two practices only shift the demand for firewood and building materials to neighboring communities, which cannot eventually solve the ecological problem in the preservation of the forests.

As regards the management of forest resources, apart from signing contracts, there also exist village regulations, which have explicit strictures against the illegal cutting of trees. Very heavy fines are imposed as punishments; for example, a fine of 1,000 yuan will be levied if a mature tree is cut down; and a person will be fined several hundred yuan if a young tree is cut down. These regulations play a vital role in the preservation of the forest and the natural environment, which also greatly benefits ecological tourism development in the local area. This, however, causes much inconvenience in the collection of firewood and pine needles by the Luoshui villagers who have to spend more on buying timber, and this poses serious problems for those families with low income.

Mosuo and Pumi peoples have similar livelihoods, production systems, and traditional customs. They also have the same methods of forest preservation and utilization of forest resources. However, because the Pumi people live on the upper slope of the mountain, and the Mosuo close to the lake shore, it is more convenient for the Pumi to collect firewood than for the Mosuo villagers. Since the development of tourism in the local area, the Mosuo people generally have more income than the Pumi, and therefore, it is more difficult for the Pumi to buy timber. The situation is far worse for the poor Pumi families. The poor Mosuo families also face greater problems than the well-off ones.

Problems Facing Luoshui Villagers

There are several serious problems regarding forest management for Luoshui villagers. First, building materials and firewood are consuming a vast quantity of wood. In the past, the matrilineal system did not split big families into smaller ones, hence people did not want to build more homes. And even if some did, the consumption of timber was not substantial. Since 1992, the villagers have begun to build more and more new houses each year, all in the traditional wooden style, which requires a great quantity of timber. Table 6.2 shows a comparison of the consumption of timber in wooden buildings of various traditional styles and modern wooden houses.

Table 6.2
Timber Consumption for House Construction

Style	No. of floors	Timber needed for the wall (m²)	Timber needed for the whole house (m²)
Traditional wooden house	1	30	50–60
	2	40	60–80
Principal rooms	1	50–60	80–90
	1	30	60–70
Modern house	2	40–60	80–90
	3	60–70	120–200
	4	100	300

Source: Yang Shoumin, a local carpenter in Jianchuan.
Notes: The principal room, traditional house, and modern house are all wooden. Modern houses do not use timber for the roof, but their roof beams and pillars are all wooden. The "timber needed for the whole house" is a rough calculation, in case of tiles, the actual consumption of timber will be less.

Although new-style wooden houses are divided into two parts without using the traditional log beams, the consumption of timber is higher by 10 m³ per floor, and the diameter is also larger than in traditional style houses. Generally, the diameters of supporting posts in the traditional wooden house are 10–15 cm, whereas in modern wooden houses it is mostly 20 cm. Besides, each household now tends to build one or two larger houses with more floors, which consumes an even greater quantity of timber. Nevertheless, at present there seem to be enough houses for the villagers, and the demand for timber is expected to decline gradually. In contrast, the demand for fuelwood is expected to remain high. According to our on-the-spot investigations, all households now use firewood for fuel. If a typical household of nine people needs 50 kg of firewood a day, the yearly consumption of wood would be 20,000 kg, which far exceeds the regeneration levels for firewood. Now that deforestation of natural forests is banned, the timber grounds will soon exhaust their supply of firewood. The villagers will then have to collect firewood in the forest, which constitutes an even greater pressure on the forest. Even with improvement in the living standards of villagers, so that more villagers can afford to use coal, electricity, and liquefied natural gas, it is difficult to change their age-old customs and habit of using firewood for fuel. In particular, people do not want to give up the "firewood hearth" which symbolizes the culture of their ethnic tribe. Likewise, they do not want to switch to coal because of the pollution. This impending fuelwood crisis remains a problem that needs an immediate solution.

A second problem for the Luoshui villagers is that while the gender division of labor tended to be equal in the past, the power of men is now being consolidated and strengthened. Although of equal labor value, women's activities and capacity in forestry are not valued or recognized. They have fewer opportunities to join in scientific and technological training in the development projects of village communities. In this regard, woman committee member, Cao Xuezhen complained, for example, that while women usually plant, weed and fertilize apple trees, it is not the women but the men who receive technical training for this activity. She further added:

The men can do nothing after training. Last April the leadership of the village was re-elected. People said that as we are a 'Kingdom of Women,' a woman should be the head of the village government, and this said, a young girl was elected village head.

She has not assumed office yet, but is working as a tourist guide in Lijiang.

Whatever it may be, the election is merely meant to show that it is justifiable to have a woman village head in a matrilineal community, but it does not mean that a woman is actually allowed to do the job.

A third problem is that the development of tourism has ushered in pornographic sex trade. Beginning in 1977, some people from outside the community opened beauty parlors and similar ventures to engage in pornographic sex trade, which seriously contaminated the village. In 1998, these beauty parlors were ordered to close down. As a result, they moved to the administrative district about 1 km away and continued their trade. This is something that should not be ignored.

Finally, the environment pollution does not allow for any optimism. With the opening of hotels and the increase in tourists, Lugo lake is being polluted, and the surface of the lake is filling in piece by piece. Administrative regulations forbidding this have been issued. Though some results have been achieved, it is very difficult to execute all the regulations completely and thoroughly.

To solve these problems, it is essential to: First, promote the villagers' perception of the importance of protecting the environment and their culture until it becomes a self-conscious choice and use the conventional regulations of the village to resist the pornographic invasion from outside. Second, attention should be paid to gender in the development projects so that more women will be able to participate, and women's abilities promoted. Last, various fuels should be used together—firewood, electricity, liquefied petroleum gas, and marsh gas—to reduce the consumption of firewood.

Zhengbo Village

Zhengbo village is situated at the edge of Yongning plateau in the lower valley of the Kaji and Dashi rivers. With Lion Hill behind it, Zhengbo faces the plateau. The village is actually a combination of three natural villages—the upper village, middle village, and lower village, which stand in a line at the foot of Lion Hill. Originally, Zhengbo was a Mosuo village. Later, Hans and Zhuang moved in successively, and each ethnic nationality lives in its respective community. At present, the inhabitants of the upper village are Mosuo and

Table 6.3
Population and Natural Resources of Zhengbo Village in 1998

	Upper Village	Middle Village	Lower Village
Population			
Total population	313	245	226
Total households	45	30	29
Women	82	74	87
Men	75	56	72
Total labor power	157	130	159
Natural Resources			
Paddy land (mu)	170	189	210
Dry land (mu)	637	701	629
Total land (mu)	807	890	839
Total rice output (kg)	90,000	118,000	1,190,000
Total maize output (kg)	34,000	50,500	44,000
Livestock			
Cow	81	56 (10)	63 (11)
Horse	80	57 (6)	66 (7)
Mule	16	37 (3)	32 (4)
Pig	486	396 (175)	341 (161)
Sheep	96	0	0
Forest Produce			
Orchard (mu)	94	92.5	85
Green thorn fruit (100 kg)	45	45	60
Mushroom (100 kg)	310	280	250
Songrong (a kind of mushroom) (100 kg)	105	93	80

Source: Comprehensive annual statistics from the Yongning county government.
Note: Figures in parentheses refer to adult animals ready for slaughter.

Zhuang, and the inhabitants of the other two villages are entirely Mosuo. The Hans have their own village, called New Village. Mosuo culture exerts influence over the other two ethnic nationalities except in matters relating to marriage and family traditions. Table 6.3 shows statistics on the population and natural resources of the three natural villages.

Zhengbo village retains its traditional agricultural structure with animal husbandry as a side occupation. Zhengbo village has fertile soil and is situated in one of the rice production zones of Yongning plateau. It abounds in rice and maize. It also grows buckwheat, highland barley, and beans. In the past, Zhengbo village grew only one crop a year, but

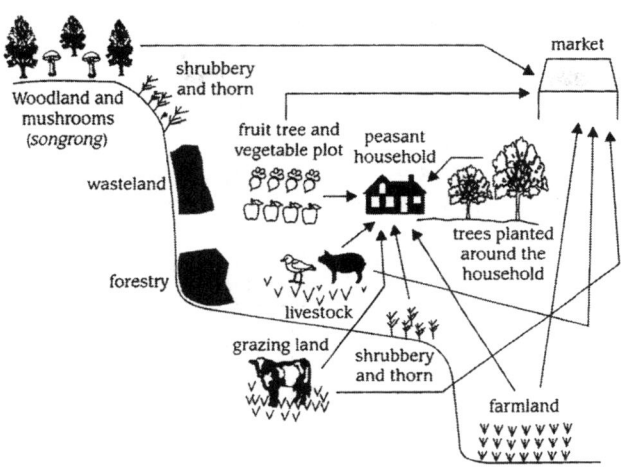

Figure 6.1
Community Resources Tendency Chart, Yongning Township, Zhengbo Village

now people are beginning to grow wheat in the winter. One-crop farming and animal husbandry do not produce much income. In some households, people have to go elsewhere for odd jobs to earn more money. They also sell some of their grain, livestock, mushrooms, and medicinal herbs in order to buy daily necessities such as woven bamboo articles, bamboo brooms and butter. Villagers commonly keep a dozen pigs, and four to five horses for home consumption or use, but seldom sell them. At most, they may sell a few chickens or piglets, and in the case of urgent need, a horse or a cow may be sold. For the direction of flow of natural resources in Zhengbo village, see Figure 6.1. Figure 6.1 shows that the natural resources of Zhengbo village have not turned out to be an advantage for commodity sales. People still rely largely on the natural economy for self-sufficiency.

Gender Relations in Zhengbo

In Zhengbo village, the relationship between the sexes is quite traditional. The division of labor remains the same: men work in the fields, and women do the housework. Women do the daily chores, though men share some of them. For instance, young men cut grass for the cows and carry it back home. Men also carry babies and do the cooking, too, which is rarely seen in other Mosuo communities. So far as the total amount of labor is concerned, there is not too much difference between sexes. However, in the slack season, women work more than the men. According to our investigation of 25 households (in the form of a questionnaire), 74 percent of the men said that the working hours of the women were longer by 30–40 percent compared to men. Our observation showed that on rainy days women were busy cutting leaves, collecting grass for pig feed, etc., while young men sat around chatting or playing *mahjong*. However, when it stopped raining, men were also busy chopping firewood, wielding flails, mending ditches, or doing woodwork, while women dried grain in the sun, cut firewood, cut grass, or did the washing. The whole village was busy doing this or that. Yet, when work outside the house is finished, women will continue to do the cooking or feed the pigs, and men will sit idle at their leisure. As the saying goes, "Get the women to look for the firewood and carry water but leave the men to do the heavy work and make money." This is a general principle in the gender division of labor. As to the control and distribution of family resources, such as the use of land and household expenditure, both women and men enjoy equal rights. The whole family decides on matters like buying large farm tools or large wares or asking for loans. This is very common in Mosuo communities.

The Zhuang mostly live in the upper village, totaling 13 households. Locally they are called the Zhongjiao tribe, which might actually be the Zhong clan. The Zhuang practice a patrilineal system, but women and men enjoy relative equality. If there is no son to continue the clan name, the family may take a man into the family as a son-in-law without his changing his original family name. That is why people often say "sell the man but not his family name." The children in this family then pick up their father's family name, and the son-in-law also has equal right of inheritance. Traditionally, there is no intermarriage among people of the same family name, and no marriage to people of other nationalities either. However, villagers began to marry people of other nationalities after the reform and opening to the outside world. At home, the most qualified member manages family affairs regardless of gender. All family members enjoy the right to own common property, and the right to control the use of the land. However, there are still some restrictions on women. For instance, when old people die, women are not allowed to come close to them, and only men can wash the corpse. The religious "master" must also be a man. In the old days, parents arranged marriages, but nowadays people may choose their partners of their own free will. Some of them are even beginning to practice the visiting marriage custom with the Mosuo.

Women Play a Major Role in Afforestation

Until 1956, ownership of the forest rested with the *tusi*, after which it was taken under state owenership. In 1961, collective ownership of a part of the forest was instituted, and in 1982 private use of part of the hills established. At present, there are only two kinds of ownership: state ownership and collective ownership. The part of Lion Hill facing Yongning plateau belongs to the collective, and the hilltop and the otherside of the hill belong to the state. The collective-owned forest of the upper village, middle village, and lower village is divided by hill ridges with distinct direction and limits. The total area of Lion Hill is 8,483.4 *mu*, of which 4,527.0 *mu* is forestland, and 3,956.4 *mu* barren hill. The forest is mostly owned by the state under the management of Yongning district forest station of Yongning district forest bureau. Forest under collective ownership is taken care of by a person sent from each village who is paid by the village out of collective funds. Besides, the administrative village designates a forest keeper to take care of the forest of each village, and supervise the work of all the hill keepers.

According to villagers Ruanku Dayu (83, woman), Awo Shidan (56, man), and Gewa Zema (69, woman), Lion Hill used to be covered by very dense forest. It was a holy hill, and so the *tusi* permitted nobody to cut trees there. People dared not cut the trees themselves. In the forest there were river deer, *muntjacs*, bears, leopards, and some other wild animals. At the foot of the hill, were very large trees. Unfortunately, this rich forest did not last. In 1947, a fire burned a lot of the forest. Some of the large trees were burnt from the top, but these later grew again. Then, from 1968 to 1970, the Kaiji river was cut into a straight channel, and thousands of people gathered here to dig. Laborers cut down numerous trees for cooking. Later, when the May 7th Cadres School was opened here, the forest suffered further damage from the demand for wood for cooking and house construction. In addition, with the cutting of firewood by the district forest bureau, not a single tree was left in the end, and the villagers were left without firewood. As a result, they dug out the tree roots for fuel. Under this circumstance, even young trees could not grow any more. Forty-four-year-old Akuruo told us, "Because no trees were available, I dug roots for firewood from when I was 14 or 15 years old until I was 26."

That was how a once very green Lion Hill turned into a barren hill. The forest in Zhengbo village was cut abusively during the time when the local people were engaged in the collectivization campaign. Neither the village committee nor the villagers could stop it because it was caused by the administrative authorities and the management system of that time. After their mountains were cleared of all trees, the villagers had no choice but to dig up the tree roots for fuelwood. The absence of an effective community management system as well as the abolishment of most community conservation traditions contributed much to the severe destruction of the forest and ecological environment. This lasted until the late 1980s.

Since 1990, the district forest bureau has strictly enforced the closing of hillsides for afforestation. It provides free saplings in July and August of each year, and each household is required to contribute one person to plant trees on the hill. From 1990 to 1998, the area of newly planted trees increased to 1,862.4 *mu*. Simultaneously, strict management has now been enforced. Besides the hill keepers and forest keepers, each village has its own forest management regulations. No living, standing or young trees may be cut. No herding is allowed in the afforestation zone either. Any violation of the regulations invites punishment. For instance, one has to pay a 5–10 yuan fine for the herding of one domestic animal, and a 50 yuan fine for cutting down a tree. Most of the villagers observe the regulations well for they suffer from a shortage of firewood, and the situation of "no water in winter but flood after flood in the summer" is a bitter experience. It was correct of Akuruo when she said, "Now everyone is aware that there will be floods when all the trees are cut. No one wants to cut trees now." Awo Shidan put it this way, "Deforestation is very bad and benefits nobody. If there is not a single tree on the hill, floods will come, which will sweep away the houses and farmland, and you won't have drinking water, either."

Nowadays, villagers are enthusiastically planting trees. Wherever we go, we can see lots of poplars and apple trees planted around the houses. There are orchards, too. The increase in the number of domestic animals which might eat the saplings has led the villagers to plant trees inside protective thorny bushes. Trees along the road are enclosed with thorny twigs and thorny bushes are planted for fences. In doing so, domestic animals are kept from spoiling the trees while afforestation has greatly beautified the fields and hills.

Women play an active role in afforestation. As far as we know, in voluntary afforestation, the participants are mostly women who plant trees around their houses. When there is a fire in the forest, it is mostly the women who rush to put it out. Hence, staff members of the district forest station were excited when talking about this. They told us that

In the afforestation, 80 percent are women; in putting out a mountain fire, 40 percent are the women; in planting trees, 80 percent are women. And when fighting a fire, it is easier to moblilize the women than the men.

Obviously, women play an important role in afforestation. In cutting firewood too, women choose to cut down the twigs, but men cut the trees at random. Men even cut down 10 cm-diameter saplings. In fact, they are destroying the forest. We saw that with our own eyes!

Problems Facing Zhengbo Villagers

The provision of firewood is an urgent problem. Firewood is in very short supply because there are no trees around the village. Besides, the villagers use only one type of fuel. In Zhengbo village, the main fuel is firewood coupled with a small amount of sunflower stems and maize cores; but the consumption of firewood is considerable. Two stoves—the cooking hearth and the pig-feed stove—daily consume 75 kg of firewood. One year's firewood necessitates a strong man and two horses to collect. In addition, as there is no firewood on Lion Hill, and as the collective-owned forest of upper village is 10 km away at the foot of Laowuji village, a return trip to collect firewood is 20 km. The middle and lower villages have no trees to cut in their collective-owned hills either. As such, they have to enter Sichuan province to cut firewood (there is an agreement between the two provinces for the exchange of raw firewood). This is also a 20-km return trip. In fact, villagers of the two provinces are cutting trees in the state-owned forest, and as a result there are almost no trees left there. In Zhengbo, we witnessed a busy scene with people rushing about to collect and transport firewood. People got up very early in the morning and left home in groups with their horses, axes, and lunch. They came back overburdened with firewood when the sun was setting. The Mosuo now say, "In the old days, the cows had the most difficult time, but now it is the turn of the horses." According to our calculations, loads equal to three trees of 10 cm-diameter can be hauled each day, and 30 days of hauling means 90 large trees cut a month. Three months of firewood harvest at the rate of 90 trees a month makes it 270 trees each year. This is the annual consumption of firewood of a household. There are 104 households in the village, and it is easy to see that consumption of firewood for the whole village is immense. In fact, many villagers now cut branches or twigs since there are, in fact, no large trees left.

Fierce floods are also a severe problem for Zhengbo villagers, located as they are in the lower valley of two large rivers. Deforestation of the upper valley causes floods in the lower valley every year. The old people recall a time when the rivers rose in summer, but it was clean water, and there were no floods at all. Nowadays, floods with mud and sand cascade downward, inundating the houses and the fields, and there is no water fit to drink. Floods have become a terrible calamity for the three villages, especially the lower village. The floods inundate two-thirds of all the fields and villagers have been appealing for a solution. In the questionnaire we presented to the villagers, both the older generation and young adults regarded floods as the second most important problem facing villagers, after money (old persons' first priority) and agriculture (young persons' first priority), as shown in Figure 6.2.

Animal husbandry also challenges the forestland. Animal husbandry is an industry of importance second only to farming. Because the grassland is often inundated and the domestic animals are short of grass for feed, this problem is becoming more serious with each passing day. People, particularly villagers of the middle and lower villages, drive their domestic animals into the forest for fodder despite village regulations. The animals destroy the young trees in the hills, and tear and eat the bark of the larger trees. They not only spoil the forestland at the foot of the hill behind the village, but also destroy its undergrowth vegetation.

Farming as a single-product economy may solve the problem of obtaining food, but it is incapable of improving the living standard of

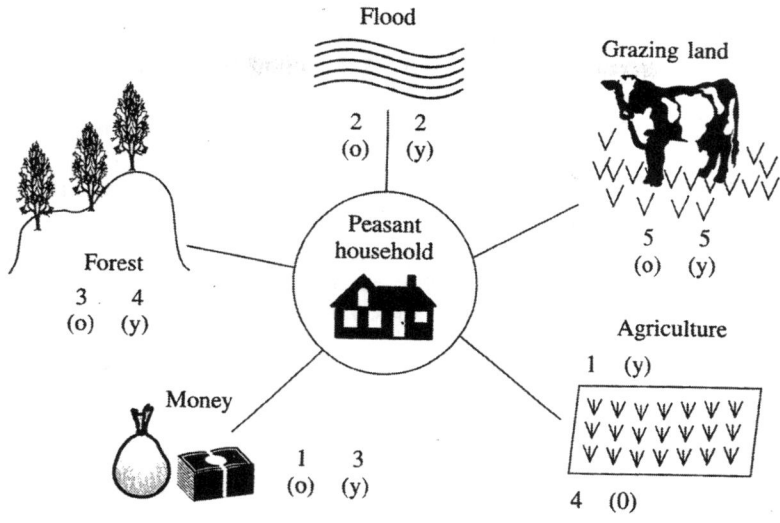

Figure 6.2
Ranking of Problems by Old (o) and Young (y) Villagers in Zhengbo

the villagers to any great extent. Until now, only three households own a washing machine or telephone, and the whole village has only one truck to haul wood and that too is not in use. Sixty percent of the households do not have a television set. In the middle village, there are only 10 television sets among 30 households, out of which eight are black and white sets. This is the typical standard of living among the villagers.

Several solutions to Zhengbo villagers' problems can be proposed. First, peasants should be given a plot of hill land for private use. Allowing them to use the barren hills and barren hillsides for planting trees will help expand the green space, and guarantee a higher survival rate for afforestation. Besides allowing hill plots for private use, villagers should be encouraged to plant trees on the appropriate barren hillsides and plateau under the overall management of the village government, by which the tree planters own the trees themselves. In this way, the villagers' enthusiasm for afforestation may be considerably enhanced. Use of liquefied petroleum gas, planting of firewood trees, and feeding pigs raw rather than cooked feed, should be popularized to regulate and save energy resources. In this way, the pressure on the forest will be eased. Forage grass to meet the needs of animal husbandry should also be improved. Put an end to soil erosion forever! We suggest that the county government lead the people in soliciting loans and securing funds so as to solve the problem of erosion from the upper valley, thus relaxing or putting an end to floods in the lower valley.

These solutions are basically concerned with the communities themselves, except for the last one, which involves outside help. Therefore, we suggest that an application be sent to the Food and Agricultural Organization of the United Nations to implement an experimental and comprehensive community development project in the three villages of Zhengbo.

Conclusions

Our investigation of Luoshui and Zhengbo villages proves the validity of our theoretical hypotheses if we limit ourselves to the perspective of the local communities studied. The practices in Luoshui village show that economic development promotes a higher living standard for villagers. When people have other channels to make money, they demand less of local forests as a source of direct livelihood, which is objectively advantageous to the protection of this forest, and which in turn promotes the development of the ecological service function of the forest by bringing about great ecological and economic benefits to villagers. A tourist industry which places value on protection of the forest reinforces its ecological service function. Accordingly, the ecology of the forest and the villagers' income enter into a virtuous circle at the local level. However, it is also true that demands for firewood and timber created by tourism are merely displaced to other nonlocal forests, just as other tourism-related social ills such as prostitution are displaced to areas outside the local community. The example set by Zhengbo village shows that a poorly developed economy forces villagers to make excessive demands of the forest, which is bound to affect the ecological function of the forest, and which in turn affects the productive and economic lives of villagers. Hence a vicious, rather than virtuous, cycle is put in place.

The stability of the division of forest ownership and policy factors are closely related to the benefits available to villagers. Luoshui village is an exceptional case. Although the ownership of the forest was taken back by the state, the villagers receive a rich return from the ecological effect of the forest because the village is situated in the natural protection zone. Therefore, the division of the ownership has little to do with their fundamental interest. However, the story with Zhengbo village is different as the division of ownership here has undergone many changes. The villagers' private ownership of hills has not been finalized until now, and there is no forest on the collective-owned hills. Worse still, the hills under their care are far away from the village. All this has directly affected production, and the life of the villagers as well. During the Cultural Revolution, rampant deforestation on the part of the government severely harmed the interests of the villagers. Once the collective-owned forest suffered from severe destruction, the state-owned forest suffered too. The trees of the collective-owned forest in Lion Hill were all cut down, and when the villagers did not have firewood they were compelled to cut the trees in the state-owned forest. Hence, the forests of both the state and the collective were destroyed. The villagers do not have any private plots to plant trees. On the other hand, they cannot be without firewood. This is a sure contradiction. Finally, it is clear that the matrilineal system, the conventional regulations of the *tusi*, and some of the religious strictures are helpful for the protection of the forest. It is advisable that the management of the forest at present make use of the scientific and effective kernels of development: gender equality in social relations, and women's adequate participation in community and forest management decisions.

REFERENCES

County Annals of the Ninglang Yi Autonomous County (1993) Compiled by the Compilation Committee of the County Annals of Ninglang Yi Autonomous County, Yunnan Province, Yunnan Nationality Publishing House.

Forest Annals of the Ninglang Yi Autonomous County (1995) Compiled by Forest Bureau and other institutions of Ninglang Yi Autonomous County, Yunnan Province, Yunnan Nationality Publishing House.

VII

Naxi Women: Protection and Management of Forests in Lijiang, China

Yang Fuquan and Xi Yuhua

Naxi villages in Lijiang county, Yunnan, China have long practiced indigenous systems of forest management which have helped preserve forest resources, in the face of the political movements of the 1950s and 1960s, and the economic changes of the 1980s. This chapter looks at forest management and women's relationship to the forest in two Naxi communities. The first (studied by Yang Fuquan) comprises the seven natural villages of the Longquan administrative village in Baisha township of Lijiang Naxi nationality autonomous county, located northwest of the Lijiang plain. Although the forests here suffered severe losses during the Great Leap Forward (1958–60), they have since regenerated under the villagers' traditional management system. Villagers here chose not to devolve responsibility over forests to households during the 1980s, as was the case with many other Naxi communities in Lijiang. Instead, the seven villages continued to jointly use and manage their forests and fields, following their traditional form of collective management and use of village forest resources. In addition, the villages continued to contribute, with less success, to the management of forest resources in the nearby *gong shan* or "public mountain"—comprised of neither state-owned forest nor village-owned collective forest, but still belonging to the eight original communities (*ba jia*) which first settled around the mountain, five of which are now included in Longquan village.

The second community considered is Enzong Naxi village (studied by Xi Yuhua), one of three natural villages in Meiquan administrative village, Lashi township, located 8 km from old Lijiang town. Forest resources of Enzong village largely survived the political upheaval during the Great Leap Forward and the Cultural Revolution (1966–76), and the opening of the market for timber and wood during the devolution of forest management in the 1980s. This was in stark contrast to neighboring villages which cut trees in their collective forests for sale, causing once lushly forested mountain slopes to became almost barren. Management measures taken by Enzong's forest management committee have so far allowed protection of over 50 percent of forest coverage in collective village forests. Enzong's traditional family system, beliefs and practices of forest worship, and various customary and cultural restrictions have been of special importance in preserving Enzong's forests.

Patterns of Forest Ownership

The community-owned collective forest in the first study site of Longquan administrative village is divided into four sections by the villagers. The forests of Songyun, Qingyun, and Hongshan villages are collectively called the Songyun Part; the forests of Zhonghe and Jiewei villages are the Jiewei Part, and the forests of Renli and Wenming villages stand alone, and are called the Renli Part and Wenming Part, respectively. This division of forests has not changed since the time of the Republic of China (1911–49). Each village is responsible for looking after its respective forest section. Since the forests of the four parts are contiguous, the boundaries between each section are marked by ravines or an obvious rock or a pile of stones, with the Chinese character for "border" cut into the rock. Before 1949, when boundary lines of the village-owned forests were delimited, each village had to report these to the local government for official records. People in a given village are not allowed to cut trees or branches, or collect firewood or dry pine needles in the forests of other villages. Community forest guards for each section are elected by the villages which share the same public forests. This is an important difference between the collective village forests and the public forests in the public mountain, which have no forest guard.

Before the 1950s, in addition to the village-owned forests, some forests belonged to private individuals who had inherited them from their ancestors. These private forests were bought from the village by rich families in early times. Some families even employed private forest guards to look after their individual forests. Several Tibetan and Yi groups in Hongshan village have forefathers who were employed by rich families to look after their private forests. In the collective forests of Wenming village, for example, there is a part of the forest called "the woods of Mr. Huang Changshou," as it had once belonged to his family. Some rich families and landlords in Wenming also owned large groves around their family graveyards which they bought from the village together with the land. The village gave these graveyard owners ownership certificates (literally called "mountain contracts"). These families also employed their own forest guards to look after family graveyards and the wooded groves around them. Other parts of the Longquan collective forests traditionally belonged to local Tibetan and Chinese Buddhist temples. For example, a part of the forests of Renli village is called "the Hill of Monks" (*heshang shan*) as it was the private forest of the Longquan Chinese Buddhist temple. Likewise, one area of forests belonging to Hongshan village is called "the Woods of Lama Monks," as it had belonged to the Guantian Tibetan Monastery.

In 1951, the Land Reform Movement began to be carried out in Lijiang county, and all private forests and lands were nationalized. The ownership of lands and private forests in Longquan was under the collective, but the owners of graveyards retained the right to bury their dead. However, the surrounding woods were considered to be collective property and graveyard owners were forbidden to cut them. All village-owned forests were allocated to respective villages according to the former division of four parts. In 1982, the government of Lijiang county started to carry out the devolution policies of the central government including delimiting "Private Hills" (distributing collective forests to households) and "Responsibility Hills" in some places. In Longquan, however, villagers elected not to distribute collective forests to households, and therefore there are no Private Hills or Responsibility Hills in the whole of Longquan Administrative Village. The ownership of the public mountain forests remained unchanged as well.

In Enzong village, forests cover 15,000 *mu* of land in the three natural villages. The collectively owned forest has been divided into three types. The first type, village-owned forest, has been strictly controlled, with no cutting of trees or collection of firewood allowed. The second type, the publicly owned San Li Qi Cun mountain near Enzong, belongs to all the seven villages and villagers are allowed to collect firewood and cut trees there. The third type is a public mountain far from Enzong village where villagers are allowed to harvest fuelwood and cut timber. However, since the distance is too great to haul back lumber within a single day, the villagers seldom go there for this purpose.

The village-owned forest is located to the northwest of the village, where the rise and fall of the mountain range looks like a tiger crouching at the back of the village, a sort of natural protective screen for Lijiang. The nearby hills have only scattered forest resources, mostly near tombs standing quietly on the mountains. Behind the mountain, however, are dense forests, covering 50 percent of the area. This forest is 2,500–3,500 m above sea level, an ideal place for growing Yunnan pine, crooked Yunnan pine, chestnut, and a small number of Yunnan fir trees. In a valley north of the mountain, there is a stretch of watershed forest protecting 10 natural springs which serves as the village water source. Behind the village, about 10 *mu* of chestnut trees blot out the sky and cover the earth, with small streams flowing quietly downward among them. A 100 *mu* of pear trees that were planted in the 1940s also grow there, full of life. The village houses are nicely framed by apple, plum, walnut, and persimmon trees. About 300 *mu* of these apple trees were planted when barren hills were contracted to villagers in the 1980s. Compared to other places in Lijiang county, the forest here has been well protected for several reasons: it is not an area designated for logging by the government, local villagers have retained their traditional customs of forest management and protection, and the village forest management committee has lived up to its responsibilities in forest conservation.

Community Forest Guards

In the Naxi language, a forest guard is called *jjuq gua*, which means a person who manages and controls a mountain (literally, a "hill"). A forest guard plays a very important role in protecting the forest resources of a community; as such, every village is very careful to elect a suitable mountain guard. Villagers elect a person who is honest and frank, and who can act with justice. Before the 1950s, some villages of Longquan such as Renli, Wenming, and Songyun specially invited Tibetan people who came from outside Lijiang to hold the post. Villagers knew that Tibetans were strangers in Lijiang and would therefore not be influenced by the complicated kinship relations of the native villagers. Moreover, since Naxi people generally believe Tibetan people to be honest and frank, they are seen as good candidates for the post, faithful in protecting the collective interests. This is partly because the villagers of Longquan have had a very long history of trade relations with the Tibetan people. Many villagers have married Tibetans, with some of the Tibetan people having settled in Longquan.

To be a forest guard is a job in which it is very easy to offend the people of the village (or villages). There is, therefore, a saying in Longquan: "Even if the fire of a forest guard's family dies out, he will be reluctant to ask for kindling (from others)." This saying reflects the isolation of a forest guard in a village. Yet, there is also a proverb which says: "The mountain gods and goddesses are always together with the mountain (forest) guards; there are mountain spirits behind a mountain guard." From this saying, one can see how important a role the forest guards play, and how important a position she or he occupies in the minds of the local people.

The villagers of Longquan have a strong consciousness, and they cooperate well for the protection of the collective forests, especially those villagers who share the common forests. The forest guards I (Yang Fuquan) interviewed all said that they were heartily supported by the villagers. Ms He Junshu, for instance, who was a forest guard for more than 20 years from the 1950s to 1970s, told me, "If I, as a woman, could be successful in managing the public mountain forests of two villages (Zhonghe and Jiewei) for a long time, it was because our people (the villagers) backed me up."

In addition to the village forest guards, the Longquan administrative village also hires seasonal mountain patrols. Their duty is mainly to prevent fires in the public mountain and collective forests in the dry season. At the time of the interview, the patrol person for Longquan was a young man from Songyun village. He got the job only after being approved by the administrative village headmen and being tested by the villagers. He has held the position for two years and is paid a monthly salary of 100 yuan by the county forestry bureau.

Salaries of forest patrol persons have varied over time. Before 1950, each household was supposed to donate 3 *shen* (a unit of dry measure of grains = 1 liter) of wheat, 3 *shen* of beans and about 3.6 *shen* of grains to his salary each year. In the 1970s patrols were paid in work points (a unit indicating the quantity of labor performed and the amount of payment earned in rural people's communes). In the 1980s they were paid 1 yuan per day, now it has increased to 6 yuan per day.

In Enzong village, the forest guards play a role similar to those of forest guards in Longquan. The guards must be strict, and in general they are candidates from poor families who have few relatives in the villages. Poverty makes them take their work seriously, especially for people who come from other places. Forest guards must appear physically large and look severe, so that people are frightened by the very sight of them. It is said there are evil spirits behind the forest guards, such that people like to threaten their children with sayings like, "Don't lie again, otherwise the forest guard will come and get you."

Community Rules Governing Forest Use and Management

Each Naxi village in Enzong and Longquan has its own traditional ways to use collective forest resources. Customary forest management practices are called *jjuq raiq* or *jjuq hal keel*, referring to the regularly organized cutting of trees in collective forests once every two to three years. This action is organized by the *laomin* or village elders, village head and forest guard. Each yearly *jjuq raiq* rotates to a different part of the forest in order to maintain the regular growth of collective forests and the ecological balance of forest cover. This kind of regular and limited cutting is an activity strictly organized by the village people themselves. It meets their needs for timber and firewood without harmful results. However, since 1988, the villages of Longquan have stopped *jjuq raiq* in order to facilitate the recovery of the collective forests which have sustained heavy damage since the end of the 1950s. At present, the collective forests of Longquan are luxuriantly green and the rate of growth of forest cover is increasing.

Traditionally, during the *jjuq raiq*, the mountain guard first writes a number at the foot of each tree to be cut. Each family then draws lots to see which trees the family is allowed to cut. Lots for large and small trees are allocated separately so that every family will get its fair share of each. The distribution of the trees is based on the household unit as a whole rather than its composition, meaning each household has one share no matter how many persons it has. Likewise, when the village needs labor contributions for public works like bridges or roads, each household is asked to offer only one person. In this way, the village maintains a balance between the distribution of benefits and the commitment of obligations among villagers.

The custom of drawing lots for trees, called *lvl yuq*, also insures fairness in other ways. If a household draws numbers for trees in distant locations, it might be entitled to cut more trees as compensation for the labor required for long distance hauling. All villagers must be very careful to cut only the trees they are allocated, to the extent that if a tree felled by a household breaks small trees around it as it falls, a fine is levied according to the size of the broken trees. During *jjuq raiq*, the activities of the village are arranged in such a way that people cut firewood together and pile it up in regular stacks. The size of each stack is then measured, a mark made upon it, and it is left to dry on the mountain. Each household later hauls away the dried firewood according to the number of lots it drew. After *jjuq raiq* is finished, the village sends some men to the mountain to look for signs of activities carried out in violation of village regulations.

During the *jjuq raiq*, the male villagers also go to the mountain to till the soil around the base of certain trees, as they believe this aids the growth of tree seeds, both naturally seeded and those planted. Villagers will normally dig up only small oak seedlings (the local people call these *shel ba*), leaving the larger seedlings to grow. These eventually become trees protecting water sources. During *jjuq raiq*, tree branches might also be harvested as compost and mixed with manure from cattle stables for use as field fertilizer.

After the annual *jjuq raiq*, villages would traditionally buy pine seeds and sow them after tilling the soil, with the forest guard usually carrying out the work. This custom is kept alive in several villages today. For example, Mr He San, the former forest guard of Renli village, sowed pine seeds years ago which have now grown into a small pine forest; the current Renli mountain guard, Mr Zhang Wu, did likewise and the village now has a pine grove. Village forest guards also prune the new pine trees for proper growth to avoid thick and stumpy trees, locally called "frog trees." The villagers consider the rainy season the best time to plant trees. As soon as the rains come, Longquan administrative village buys tree seeds for all the villages of Longquan, and then asks each household to provide a person to plant them in the collective forests, barren hills or fallow areas. If a tree planted by a household is kept alive, the household is awarded a certain sum of money. Each household is asked to take responsibility for the successful growth of the trees it plants, and the village authority inspects the results in the next rainy season. The survival rate of the trees planted in this way by Renli, Songyun and Wenming villages has now reached 75 percent.

In the past, each Longquan village customarily closed the hillside to facilitate afforestation at a certain time every year, usually from Pure Brightness (the 5th of the 24 solar terms) to the end of September when the rainy season is over. The local villagers believe this period to be the best time for growing trees and other plants. It is, however, not considered a good time to cut trees or leaves for barnyard compost, as it is believed that doing so will cause severe thunder and hailstorms. As a result, villagers are allowed only to gather some dry branches and leaves during this time. Picking the fruit of economic wild plants like pine nuts is also forbidden at certain times. Once ripe, the village committee will arrange an exact time to pick pine nuts and other fruits in the collective forest. This custom is still followed in villages like Jiewei and Songyun.

The villagers of Longquan are allowed to gather dry pine needles in the collective forests and the families which are not able to afford tiles for house walls are allowed to cut pine tree branches to cover the tops of walls. In this traditional method of preserving house walls, soil is pressed over the pine branches. Today, Longquan administrative village and the other villages of Longquan all have their own village rules. For example, the 11th regulation of Qingyun village is: if a fire breaks out, whether in the village-owned or state-owned forests, all villagers have to participate in putting it out, and people who do not help will be fined 10–30 yuan. The 10th regulation of Renli village includes such rules as: If a person cuts down a green pine tree which can be used as a rafter, they will be fined 100 yuan; if it can be used as a roof beam, the fine is 200 yuan; if it is larger than this, the fine is 500 yuan. The guilty person must pay the fine within three days, otherwise he will be punished by decision of a village meeting.

Enzong village, like Longquan, has clear rules on the use of the collective forest, and clear consequences for its misuse. In January

1995 these rules were stipulated by the Enzong village committee as follows:

1. Except during permitted times no one is allowed to bring axes, knives or hoes into the collective forest to cut trees or dig tree roots. If they do so, they will be fined 5–10 yuan.
2. During the periods when the mountain is closed for afforestation, no one is allowed to cut fuelwood trees, dead trees or knarled wood. Those who violate the rules will be punished by fines of 5–10 yuan.
3. Anyone who secretly cuts trees belonging to the village at night will have their tools confiscated, and will be fined 10 yuan for invading 1–3 *cun* of land area, 15 yuan for 5–8 *cun* of land, and 20 yuan for 8–20 *cun*.
4. Use of baskets is allowed for carrying pine needles, but they cannot be used to carry cut young trees or branches. Those who violate the rule will be fined 3 yuan for young trees, and 3 yuan for tree branches.
5. Cutting tree branches is not allowed. Those who violate the rule will be fined 10 yuan each time they violate the rule.

In total, there are 12 articles in the forest management rules, and 10 duties for the forest guard stipulated by the village committee. After negotiations between the committee and the forest guard, an agreement stating the rules is signed by both parties. In practice, these measures have turned out to be very effective, and the instances of secret cutting of trees is now clearly less than before.

Managing the Public Mountain

In addition to collective forests, Longquan villages have a nearby *gong shan* or "public mountain" which has belonged to the eight founding villages since ancient times. This mountain is formally called *shu he ba jia gong shan*, literally "the public mountain (forests) of the *ba jia* (eight villages) of *Shu He* (the old name for Longquan)." At present, the mountain still belongs to the original eight villages and is unique in that it is a "public mountain" which is neither part of state-owned forests nor village-owned collective forests, and has its own management system.

In the past, the residents of the eight villages were allowed to collect firewood and cut down trees to build houses as well as herd domestic animals on the mountain. Although the timber taken from the public mountain was essentially free, there were also established traditional rules of its use and management. Each village generally only cut down trees, collected firewood, and cut tree branches in the area of the public mountain close to the village. But this custom was broken during the Cultural Revolution and people of different villages cut trees everywhere on the mountain. Since there is no forest guard for this public mountain, the eight villages have a difficult time managing it, especially because it borders on the hills of the other administrative villages and townships. At present, although the trees on the mountain are mainly cut by the traditional eight villages, sometimes outside villagers also log on the mountain, without supervision.

According to Mr Li Wenyong, who was the officer in charge of forests affairs for the Baisha township from 1984–90, before the 1950s, unregulated cutting of trees on the *ba jia* public mountain was not allowed. When Li Wenyong was a boy, many trees big enough to be used as house beams remained on the mountain. Mr Yang Peicheng, a local retired teacher, and a man who is very familiar with the history, culture, and customs of Longquan, told me (Yang Fuquan) that the *ba jia* villages had an unwritten rule: villagers who cultivate crops or herd livestock on the public mountain have the obligation to protect the surrounding trees. The villages which formerly belonged to *ba jia,* but are now under a different administrative jurisdiction still abide by this rule. Before the 1950s, the eight villages had customarily contracted management of the public mountain to the households. If a household planted some crops (like potatoes) and herded its domestic animals on the mountain, it would not only be responsible for protecting the large trees in the area, but also had to hand over some materials as a substitute for tax on the mountain forests.

Traditional Spirit Beliefs Protecting Forests

Traditional beliefs are also an important factor in the management of Naxi forests and water resources. In the past, in addition to worshiping Shu nature spirits, the villagers of Longquan and Enzong also paid homage to mountain gods and goddesses, and dragon kings. They believed that the mountain spirits controlled the forests, water sources, and wild animals. Every family would first offer sacrifices to the mountain spirits when it went to visit family graves to honor ancestors. The trees and bushes beside the water sources are also considered to be the dwelling places of the mountain spirits and the dragon kings, and it is strictly forbidden to damage even one small part of them. Longquan villagers believe that the ponds of Jiu Ding Long Tan (Dragon Pond) and Po Ding Xiao Tan are sacred pools where dragons reside. As indications of this, dense trees surround the ponds and special fish with protruding eyes inhabit them. In *The Historic Record of the Qianlong Period* (1736–95) of the Qing Dynasty, there is a saying:

‹ The Ai'a Pond (Jiu Ding Long Tan) is 10 Li (half a kilometer) away from the town (of Lijiang), its water is limpid and fish abundant. The size of the pond is about 10 *mu*. The rocks surrounding it are so beautiful and the trees so dense. It is said that when a certain person once tried to catch fish in the pond with a net, it suddenly began thundering and lightning, and since then, no one dares to fish in the pond.

Such customary taboos worked together with the various village regulations and rules to control and guide villagers to treat the environment in the proper manner and use forest resources carefully.

In Enzong, in the third month of the Lunar year the "Sacrifice to Shu" ceremony is held. Shu represents spirits of natural phenomenon like mountains and water. It is said that originally, humans and Shu were brothers of the same father, but a different mother. It was humans who destroyed nature, so Shu took revenge against humans by causing natural disasters. Later on, humans and Shu reached an agreement that humans were allowed to cut trees and hunt wild animals in small amounts. Humans ought to offer sacrifices of food to Shu in March of each year, feed Shu with women's milk to cure the wounds caused by humans, pray to Shu to protect them and bring good weather, and ask Shu to forgive their mistakes in cutting trees. During this ceremony, no blood would be seen, and once the ceremony was completed, two live chickens were offered to Shu to pay for the debt of cutting trees and hunting wild animals. On 29 or 30 July the Jia Ri Duo Fair also took place where a Lama was invited to chant scriptures, give medicine to the dragon king, and scatter pure water into the springs. A white cloth printed with sacred scriptures was fixed to a pole and lodged on a mountain top behind the village, illustrating the gratitude of the villagers to the mountain god.

Impact of Social and Policy Factors on Forests

From the Land Reforms to the Cooperative Transformation of Agriculture (1951–57), the central and provincial governments emphasized that "protection of forests is more important than afforestation," and organizations for protecting forests in rural areas were established. There were very few activities harmful to forests in Naxi communities at that time. The traditional Longquan village rules and regulations for protecting the collective forests were compatible with the policy and the forests were very well preserved. But in 1958, the Great Leap Forward started throughout China, and the collective forests were struck by misfortune. At that time, cobblers and blacksmith cooperatives and factories for producing Chinese "gold thread" rhizome drugs were established in Longquan villages. Many cobblers and blacksmiths from different places were gathered in Longquan, a public dining room was set up, and villagers were organized to smelt iron and make charcoal on a large scale. All these activities required a lot of wood and thus a mass of people ran into the collective forests to cut down trees and firewood. As a result, many big pine trees and dense oaks were cut down. For example, several hundred huge oaks in the Hill of Monks behind the Longquan temple were all cut down. After 1958, the production teams of Longquan built a granary and drying grounds for grain storage, and cut many large trees to make the grain racks. At the same time, the production teams broke rocks with explosives to build the granary and the grounds. What was the result of these activities? Not only was the forest vegetation seriously damaged, but the water source of the two ponds (Jiu Ding Long Tan and Po Di Xiao Tan) was also affected, and less water than usual was produced. Mr Yang Peicheng told me that when he was young, the water literally gushed out from the spring of Jiu Ding Dragon Pond, but later the water flow grew weaker. According to Ms He Junshu, who was mountain guard for more than 20 years, when she was a girl, the two ponds were very large and wide, and there were many streams flowing along the foot of the hills, but now most of them have dried up.

In Enzong village, the Great Leap Forward was a time of organizing public kitchens, and building the barn and frames to dry corn for the collective. Cutting timber in the name of the collective required no special permission, and many mature trees were harvested just for fuelwood. Since the public kitchen burned fuelwood in addition to that consumed by private households, demands for fuelwood were even greater. Even chestnut trees near water sources were cut. Before the 1950s and much earlier, local Naxi people had been warned not to cut the trees near the water sources, believing it would bring disaster and insult the mountain and water spirits. Even today many people share this belief. In 1958, during the time of steel and copper smelting, Enzong was fortunately not chosen as a main site. Yet, someone foolishly proposed that they could burn wood to make charcoal to help with steel and copper smelting. As a result, more than 1,000 young pine trees east of the village were cut, put into a hole in the earth, and burned. But after several days had passed no charcoal was successfully produced. During the Cultural Revolution, all management measures were in chaos, the moral standard of the public had degenerated, and people looked for quick benefits. Ordinarily they would not collect firewood from distant mountains, but once they learned that someone had dug out a tree root for fuelwood and had not been punished, others followed suit. The result was deforestation and soil erosion causing once clear spring waters to turn muddy, and water sources and springs to dry up.

After serious deforestation during the Great Leap Forward, for quite a long period before the 1980s, the collective forests of Longquan were not degraded on any grand scale, but selectively cut under the annual *jjuq raiq* arranged by the villages. Collective forests were under the able management and control of the villages. But from the late 1980s, especially from 1985–88, many Longquan villagers were affected by a storm of illegal deforestation sweeping through Lijiang at that time. As the timber market was opened with the withdrawal of nine southern provinces of China from the system of unified distribution of timber, many people thought that if one wanted to get rich, they just had to go to the mountains and cut timber. Hence, deforestation became severe in many parts of Lijiang, and the state-owned and collective forests were seriously affected. According to statistics reported in 1988, the wood storage of trees dropped from 48,031,400 m^3 in 1959, to 37,739,000 m^3 in 1988, so that during a period of 29 years, 10,292,400 m^3 of trees disappeared. The destruction of collective forests and the public mountain also brought disaster to the wild animal population. According to several old people of Longquan, in the past there were several species of wild animals in the hills of Longquan, like wild boars, leopards, wolf, *muntjacs* (small deer), gorals (goat antelope), otters, and badgers. What remains today are only a very few kind of animals like rabbits, pheasants, and *muntjacs*.

After the 1950s, religious beliefs were disallowed, and with the abolishment of superstitious activities, many Enzong villagers were free from the threat of the mountain god. With villagers now lacking moral guidance, secret logging took place constantly. Any relaxation on part of the mountain guard meant someone would seize the chance to secretly cut trees without punishment. Many people saw the forest as a gift from their ancestors, and believed they had the right to use it as they pleased. As soon as this thinking was formed in their minds, they started cutting trees secretly. Logging accelerated and quickly exceeded regulations, and some part of the mountains became barren.

The village management of the collective the forests of Longquan lost control of the forests during the late 1980s. Some villagers no longer paid attention to customary village rules and cut down many trees in the collective forests. A serious conflict between villages was caused; for instance, in one year during this period, some young people of Wenming village not only cut down many trees in the collective forests of their village, but also cut down trees belonging to the neighboring Zhonghe and Jiewei villages. They cut trees through the day and night, and hauled logs back to their homes with ox carts. When some of the women of the Zhonghe and Jiewei villages who had gone to gather dry pine needles in the forest caught the young people in the act of cutting trees and tried to stop them, they ridiculed the women and cursed them. This act of aggression enraged the people of Zhonghe and Jiewei villages, and the people of the two villages took joint action to stop the violations. First, they called on all the village families to send one male member, or if there was no man, to send a woman. When several hundred people were assembled, they went *en masse* to the mountain to catch the men who were still felling the trees. They tied the two young men, who still would not admit to any wrongdoing, to the willow trees beside the stone bridge of Shu He market. The crowd also captured their ox and then went house to house in Wenming village seizing all logs which looked like they had been recently cut, and piled them up on the granary grounds of Zhonghe village. The people of Zhonghe and Jiewei villages then distributed some of the logs by drawing lots, and some logs were sold to the villagers.

Illegal logging in the collective forests of the two villages at this time had become very serious, and police from Baisha township who had previously been called to put a stop to it, could not effectively do so. According to two elders of Jiewei village, Yang Peicheng and He Junshu, many people from Wenming village thought it was appropriate that Jiewei and Zhonghe villages punished the villagers of Wenming who cut trees everywhere. They believed that the collective forests of Wenming were also being destroyed by the people who ignored the village rules, but since most people of Wenming village had kinship ties or other close relationships with them, it was difficult to punish them forcefully. While the mass action of the villagers of Jiewei and Zhonghe was rather extreme, without it, the deforestation would have continued.

Women's Role in Management of Forests

In Longquan, women have a very close relationship with the forests. Longquan was very famous for leather products in Yunnan, Sichuan, and Tibet, and many village men were engaged in leather production. Before the 1950s, women also actively helped the men in leather-working, with many of the women making laces for leather shoes. The laces made by Longquan women were known far and wide for their high quality, and therefore had many customers in Lijiang, Zhongdian, Deqin, and Heqing counties. Even the famous shoes factories in Baoshan, Yunnan, only used the laces made by the Longquan women. Since many Longquan women derived a good income from this traditional sideline production, not many of them collected firewood for sale, and most gathered enough for daily use.

After the 1950s, however, the household-based leather crafting industry declined when agricultural and handicraft reforms were implemented. People were now asked to put their efforts into agricultural work, mainly under cooperatives in urban areas. This caused a shortage of income for village households, thus making the fetching and selling of firewood a very important means for Longquan women to earn an income. The women usually went to the public mountain to collect firewood. Ms He Junshu, who was 82 years old in 1999, reported that at that time, the women of the village went to the mountain at the crack of dawn to collect firewood, returning only in the afternoon. Firewood was usually sold in the old town of Lijiang. He Junshu sometimes sold one bundle of firewood, and sometimes the two, but if she wanted to sell two bundles of firewood, she had to carry the first one to town, and as soon as it was sold, come back in a hurry to carry the second bundle for sale; very hard work as she tells it.

In 1958 when "collective dining rooms" were set up, the number of people who cut firewood increased since local supply and marketing cooperatives as well as state-owned restaurants were buying as much firewood as they could get their hands on. The women of Longquan went to the hills to collect large amounts of firewood and sold them to these work units. At that time, the Longquan market was full of huge piles of firewood which had been repeatedly carried back by the women from the public mountain. Owing to the decline of traditional handicrafts and few opportunities for sideline work, many women of Longquan had no choice but to collect firewood and sell it for income. They were usually busy all the time cutting wood, at the same time having to do all the farming work as well as all the household work. Some women who had experienced this kind of life told me (Yang Fuquan) that during the winter and summer seasons, they had to get up very early to go and cut firewood, and after returning from the hills, they had to immediately take part in collective production labor, otherwise they would have their daily workpoints deducted (a unit indicating the quantity and quality of labor performed, and the amount of payment earned, in rural people's communes). He Junshu reported that she once cut firewood continuously for more than 50 days, since at that time, firewood was almost the only income resource for her family.

The heavy cutting of firewood caused by limited income resources not only put heavy labor demands on women's shoulders, but was also a disaster for the forests of the public mountain. A lot of oak trees and various other trees including many mature pine trees were cut down. Firewood which formerly was used mostly for cooking became the main resource for the villagers' income. This change caused heavy losses of forests, and destroyed the normal growth cycle of the trees in the hills.

An indication of women's importance in Longquan is that the village had a woman forest guard, which is very rare in Lijiang. He Junshu (mentioned earlier), is a woman from Jiewei village who has performed the duties of forest guard for the collective forests of Jiewei and Zhonghe villages from 1957, when she was 40 years old, up to 1979. During the 22 years of her career as a mountain guard, she consistently fulfilled her role, even when violators of forest protection rules were very strong men or when they dominated by sheer force of numbers. Regardless, she would confiscate their axes or machetes according to the village rules, and force these proud men to make a formal apology to her or to the village committee (at the time it was a committee of the production team) and they were punished accordingly. She emphasized one point when she summarized her years of successful experience as forest guard: "My success depended on strong support from the people of the two villages, they were always behind me. The village rules were a deterrent force to anyone who wanted to cut trees illegally." Hence, when she went on patrol in the hills or guarded a road alone under the moonlight, where the men who cut trees illegally must pass, she never felt lonely or afraid. According to traditional beliefs, as noted earlier, the mountain gods and goddess are always behind the mountain guards. Yet, the power which boosted her courage the most came from the villagers. In turn the production teams and villagers also trusted her immensely and gave her nine workpoints per day, the highest possible number at that time. The other male forest guards and villagers said, "A Xian (Junshu's nickname) is a courageous and robust Naxi woman. When she was the forest guard, many men were afraid of her and also respected her."

Naxi women are the main laborers in cutting firewood and gathering dry pine needles, and therefore their suggestions are very important in decisions related to forest resources, such as the election and evaluation of forest guards. Regeneration and sustainable use of community fuelwood forests depends mostly on women's appropriate methods of firewood management. As mentioned earlier, Longquan women traditionally cut great quantities of firewood and caused a reduction in fuel resources. Old-style cooking stoves and traditional methods of feeding domestic animals were also key reasons for the high consumption of firewood. In the past, Longquan women used an open hearth style stove, called "the tiger kitchen range" by local people. Pigs were fed with cooked food which required a lot of firewood as well, so that a woman had to cut 40–50 bundles of firewood (each bundle about 50–60 kg) per year. At present, all Longquan villagers use improved kitchen stoves, and some households have even installed a hot water tank inside the kitchen range, thereby reducing the amount of firewood used, which would otherwise have been needed to heat water. In addition, many villagers now feed pigs with uncooked food, which also saves a lot of firewood. Moreover, according to Mr Li Wenyong, the secretary of Longquan administrative village Communist Party, 95 percent of the residents presently have electric stoves, and some villagers have started to use liquefied petroleum gas stoves as well. For instance, in Zhonghe village, more than 20 percent of the households use liquefied petroleum gas stoves, and some villagers have even begun to use solar energy devices to heat water.

In addition to fuelwood, women commonly use cornstalks for cooking. According to my (Yang Fuquan's) survey of 30 households in Renli village, for example, each household farm generated 500–800 kg of corn stalks each year which could be used for cooking. Stalks could be kept for several months and households thus saved a huge amount of firewood. Along with this decrease of firewood use, has come a reduction in women's labor in firewood collection and cutting so that Longquan women can engage in more household economic activities. The same Renli survey shows that the distance needed to travel to collect firewood from the public mountain has not grown any longer in recent years, and trees supplying fuelwood remain in good quantity. This is in contrast to other places in Lijiang where fuel trees (like the various species of oak trees) on public mountains have suffered from over-cutting.

As more and more Longquan village men have taken sideline jobs, most household incomes have increased, and some women have been freed from the hard life of selling firewood. In Zhonghe village, for example, many men earn money slaughtering animals in the town and their wives assist them in selling the meat, hence there are now very few women in the village involved in collecting and selling firewood. The villagers earn a good deal of money from sideline occupations, and some of them have even begun to buy firewood from others. Some men have also organized a team to cut and carry firewood with horses. They usually cut firewood on the public mountain and sell it to the people of the villages. Every village has some households with higher income, and they no longer cut firewood. Moreover, owing to the decrease in firewood consumption, stacks of old dry firewood are now kept for years by many families.

Some Lijiang factory enterprises which were heavy environmental polluters have now been shut down for rectification, and this has also reduced the burden and hardship of women's labor. For instance, Lijiang County Paper Mill stopped paper production in 1998, and the great quantity of wheat and rice stalks which were collected by women as raw materials for paper making are no longer sold to the mill. Instead, Longquan villagers now use them for barnyard compost, so that the number of times women must go to gather dry pine needles and leaves is greatly reduced. This is also very good for the recovery of mountain vegetation. The use of chemical fertilizers has also decreased the amount of labor by women in collecting dry pine needles and leaves for compost. A bundle of dry pine needles carried back by a woman can make five to six baskets of barnyard manure.

Though Longquan villages do not have many fields, at the time when all fields only used barnyard manure, a minimum of 50–60 baskets of barnyard manure was needed for every field, so each woman had to gather a huge amount of dry pine needles and leaves. Households with strong laborers often used 70–80 baskets of manure per field. However, while the use of chemical fertilizer has reduced women's labor, the use of too much chemical fertilizer has caused the soil to harden, making weeding, which is done by women, very difficult and has added to their labor.

The women of the Longquan villages, in addition to their labor in firewood, pine needle, and leaf collection, are also the main collectors of mushrooms and ferns, a source of income for many villagers. According to 1999 statistics, the Longquan administrative village has a total income of 12,000 yuan from the collection of wild mushrooms and ferns. Renli village's share of this is 2,970 yuan, Songyuan's 2,410 yuan, and Wenming's 1,900 yuan. From the late 1950s to the 1970s, owing to low production of grain, many villagers dug out fern roots as substitutes for grain in making alcohol. As a result, the number of wild ferns decreased. Nowadays, villagers have much more grain and do not need to collect ferns anymore to make wine. Moreover, decreased deforestation has improved the natural environment of wild plants, and ferns have grown again on a large scale, and have become an important income resource for the villagers.

In Enzong, it was traditionally the woman who was in charge of family affairs, and there were few other means of making a cash income outside the women's collection and sale of firewood. Before the 1950s it took women three hours to carry a bundle of firewood back home, and another half a day to carry it to the market, where it could be exchanged for 1–2 *jin* of salt. In the past, women could collect three bundles of firewood in a day, but now only one bundle is a possibility. Each year women spent a period of two months collecting and carrying pine needles, and a 100 bundles of pine needles were always ensured in the harvest in autumn. Nowadays a woman can collect only one bundle of pine needles in a single morning. With improvement in cooking stoves, especially the popularization of gas stoves, firewood demand has been reduced, but during the six months of winter, it is still necessary to burn wood to keep warm, especially for the elderly.

Today, the increase of income-earning opportunities for Longquan villagers has reduced women's hard labor of cutting firewood in the hills, and hauling to town to sell. Yet, because many men go to the urban areas to do sideline work, women now have much more work to do on their homesteads. They have to do the farming work, raise the pigs, gather needles and leaves, cut firewood, and look after elderly parents-in-law and children. In the Renli village survey of 30 households, 100 percent of men acknowledged that women were busy in both farming and household duties, and estimated their working time to be 10–20 percent more than the men. However, women in the same households commonly estimated their work time at 20–50 percent more than that of the men. Many men said that a woman is the backbone of a family; if a household had no woman to manage it, it would be very difficult for the family to be prosperous. The village headmen told me (Yang Fuquan) that they often tried to persuade single men who have difficulties in family life to find a wife as soon as possible. From such an example, one can see how important women are for family life and farming. Of course, one can also easily see how hard women work in daily life.

In Longquan villages, generally speaking, the decision-making for family affairs is carried out by both women and men. According to my (Yang Fuquan's) investigation in the villages, family affairs are usually decided by both the women and men heads of households for example, what to buy? what to invest in? Women not only play a very important role in the management of community mountain resources (although women members of the village committee are fewer than men members), but also have quite a powerful voice in the social affairs of a village since they have very important productive and reproductive roles in households, as mentioned earlier.

In contrast, in the male-dominated society of Enzong village, the women have a subordinate role in forest management. In Enzong, women have no rights to the forest, the land is inherited from father to son, forest resource distribution is done on the basis of households headed by men, whose sons have the right to apply to cut timber. Only after a woman marries into the husband's home does she have forest rights through him. Yet the hardship or leisure of women is in direct proportion to the availability of the forest. In places rich in forest resources, women have more leisure time. Forest resources are closely related to water resources, and many women select their husbands partly on the basis of whether the men's homes are in areas with sufficient forest and water resources. Though women have no direct rights to forest resources, they share rights in the name of their husbands, and greatly influence forest distribution plans.

Activities of Enzong Naxi women, like collecting firewood or pine needles are seasonal, and are social activities connected to cultural communication. The proper time to go to the mountain is normally winter. Women in the village have always liked to go together in groups to collect firewood, especially after the land was contracted to private households, and collective production activities devolved to individual households. In harvesting forest resources together, women exchange knowledge of forest protection and develop an understanding of the importance of taking care of trees. Groups of village women like this are very important sources of cultural influence and unity for women. This kind of cultural influence is crucial since women are outsiders who marry into the village from different cultural backgrounds, and it takes time to socialize them in village attitudes and values of forest management. Conversely, men who are born and brought up in the village are educated in a shared traditional culture.

Prospects of Forest Management

If we look into traditional spiritual beliefs in the management of collective forests, we find that women spirits, like Naxi women themselves, have occupied a central role in forest protection. Naxi people in Longquan, for example, believed that it was a female goddess of the earth who was in charge of the mountains. Among the spirits who control nature, there are many who are considered to be female. Traditional Naxi forest worship involves sacrifice to the god Shu, as mentioned earlier, and Shu is historically believed to be female-dominant. Shu is the spirit who controls mountains, rivers, flowers and plants, and all of nature including the places where humans reside. In Enzong, villagers believed it necessary to hold a sacrificial ceremony to Shu each year to protect the natural surroundings and maintain harmonious relation with nature. Shu is considered to be alive and aware like human beings, therefore cutting forests hurts her, and her wounds must be cured through a ceremony called "giving medicine to Shu." If a villager cuts trees near a water source or on a mountain closed for reforestation and becomes ill, villagers believe her or his spirit has been taken away by Shu in punishment. A ceremony then needs to be held to ask for the return of the spirit. *Family and relatives take sacrificial goods and some clothes from the* sick person to the side of the water source pond and ask for the spirit to return. After the sacrificial ceremony, a tree branch has to be inserted into the tree which was cut, symbolizing that the person who has done wrong gives life back to the tree.

In Naxi Dongba pictographs, Shu is written as a creature with the head of a frog and tail of a snake. In traditional Naxi culture, the frog is an image for woman and the snake for man. Shu appears as a combination of female and male, with the frog as the main body and snake as subordinate. That men appear in a subordinate position represents the historical trace of Naxi worship of female gods. From the viewpoint that religious beliefs reflect social phenomena, the worship of female gods reflects women's past predominance over men. This is further illustrated in Naxi fairy tales, religious ceremonies, and in the structure of the Naxi language. For example, in divination ceremonies, the god is a woman named Pan Zhu Sha Mei, and the sacrifice to heaven ceremony is held mainly for the woman ancestor Cen Hong Bao Bai Min. In terms of Naxi language, in many of the compound pictographs concerning female and male, words representing female are placed before the male ones, connoting power, priority, and respect. Dongba pictographs record the image of the female as heroic as well.

It was in the early 18th century, when the governing Han Chinese administrative reforms forced the local people to change traditional customs, that the predominance of women was subsumed by the male-centered Confucian cultural system in Naxi areas. In 1723, the system of traditional inherited local governors or *tusi* were replaced with governors appointed by the central government. An education system promoting men's domination was established, Han Confucian schools were opened to Naxi men who were then socialized into Han culture, arranged marriages were popularized, and traditional cremation of the dead was replaced with burials. Thus, the customary saying about the prowess of Naxi women—"once a woman enters a fight, it will be over"—lost its meaning. Openly celebrated religious activities also disappeared for a long time, but were continued secretly by groups of women. When women cut trees on the mountain, for example, they still followed the belief that chestnut trees around sacred water sources should not be touched.

Men have taken over the social position of women in Naxi society, and women now have to put their wisdom and energy into family affairs. Women's role in maintaining traditional Naxi culture and the family union is important, but is also a restriction to them.

For instance, women will express their viewpoints on village forest distribution plans, but since they live in a male-centered society, there is little possibility of them being involved in decision-making stages, and they have no rights to forests once distributed.

When abolishing traditional superstitious beliefs and activities which promote good moral standards in forest use and conservation, some equally effective system of monitoring, punishment, and fines ought to be established. In the past, forest protectors were supported by food grains contributed by each household, therefore they faced careful monitoring by all villagers. The post of forest protector was a post for poor persons who needed to perform it well in order to survive. Now it is a job paid for by the local government, with many contenders for the post. Villagers' monitoring of the forest protector is thus more relaxed, and she or he is less effective in protecting forests.

In the Naxi villages of Longquan, villagers have stopped the traditional annual practice of cutting collective forests (jjuq raiq), and most villagers now have to go to the public mountain to cut firewood. Due to a lack of community unity in the management of the public mountain in recent years, disordered deforestation of mature timber has occurred on the mountain. People from some poor neighboring villages near Longquan, which do not belong to the traditional eight communities (ba jia), have also cut large quantities of trees and firewood on the mountain. For instance, the main income resource of Wenhai village, which is listed as one of the poorest villages in Lijiang county, is charcoal-making, for which the villagers require a great quantity of firewood. Hence many of them engage in heavy cutting of oak trees on the public mountain. Some villagers of Lashi township also cut firewood on these.

In the last half century, traditional Naxi systems of community self-management and control of forests have been eroded in great part due to the serious impact of frequent political movements before 1976, especially the Cultural Revolution. Cultural factors in forest management have lost their relevance, for example, local people once worshipped mountain spirits and had taboos related to forest protection. But after 1949, such a belief system has been considered "superstition," and more and more young people ignore traditional beliefs related to forest conservation. If the public mountain remains without proper community management, and lacks proper coordination and control among villages, it will be difficult to preserve the forests there. If it stays deforested for a long time, soil erosion will certainly result and the tree-growing lands and farm fields of the villages of Longquan located below the public mountain will be affected. The turbulent flooding of the Qinglong river in recent years is an indication of what may happen with further deforestation and degradation of vegetation on the public mountain.

All Naxi women of Longquan and Enzong shoulder both the heavy loads of farmwork and household duties. Their labor in firewood cutting and dry pine needle gathering has been reduced with alternative sources of fuel and compost material, but if the public mountain were again managed properly by resuming the methods and traditions of regular forest cutting in certain areas and certain times in rotation, this would give women additional forest resources for their daily needs. Only after all of those matters have been solved, will the protection and preservation of village-owned forests be meaningful.

VIII

Bobolizan, Forests and Gender Relations in Sabah, Malaysia

PAUL PORODONG

The Rungus of Sabah

In literary writings, the Rungus are called the Rungus Dusun. While it is right to include the Rungus under the Dusunic language family, the Rungus never regarded themselves as Dusun. They prefer to be called Rungus or Momogun. In rare cases, the Rungus will identify or affiliate with Dusun or Kadazan when dealing with non-Dusunic people. Their social organization is cognatic. The major social groupings are the domestic family, the longhouse, the village, and the territory or reserve area (*pomogunan*). The domestic family is the major unit of production, consumption and asset accumulation (Appell, 1976). Surpluses from the family's swiddens and livestock production are converted into brassware, gongs, and ceramic wares, which form their major assets.

A family's rights over swidden areas are recognized until the last crop planted by them is harvested; then the swidden reverts to the area at the disposal of the village. As the jungle grows, other families in the village may cut the old swidden area again, and no rights lie with the previous cultivators. Thus, the village owns a territory which residents of the village may cultivate, but residents of other villages may not. This territory has well-recognized boundaries such as hill ridges, streams, rivers, trees, and other physical landmarks. It can be concluded that the village is the major political unit of the Rungus society (Appell, 1966).

A village may consist of one or more longhouses. Each domestic family occupies a separately owned apartment in one of these longhouses. The major feature of the longhouse social organization is the complex web of social relations in which the domestic families of the longhouse are enmeshed, and which is based on ties of kinship and mutual assistance. Neither the longhouse nor the village can be considered kinship units, for kinship is not a distinction made in the recruitment to such units. Any family can join the unit if it has the village head's (*osukod*) approval. This approval is based solely on the personality of the members of the family seeking to join.

The longhouse consists of closed and open individual compartments (Figure 8.1). The individual compartment consists of two major areas—the *apad* (open gallery, marked as A, B in Figure 8.1)

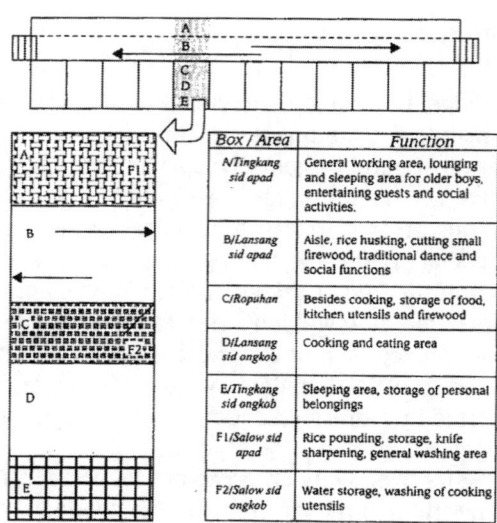

Box / Area	Function
A/Tingkang sid apad	General working area, lounging and sleeping area for older boys, entertaining guests and social activities.
B/Lansang sid apad	Aisle, rice husking, cutting small firewood, traditional dance and social functions
C/Ropuhan	Besides cooking, storage of food, kitchen utensils and firewood
D/Lansang sid ongkob	Cooking and eating area
E/Tingkang sid ongkob	Sleeping area, storage of personal belongings
F1/Salow sid apad	Rice pounding, storage, knife sharpening, general washing area
F2/Salow sid ongkob	Water storage, washing of cooking utensils

Figure 8.1
Plan of a Traditional Longhouse in Rungus and Detailed Plan of a Single Unit

and the *ongkob* (room, marked as C, D, E in Figure 8.1). The *apad* is demarcated into *tingkang sid apad* (general working area), *salow* (water use area) and *lansang sid apad* (aisle). The *ongkob* compartment is divided into *ropuhan* (the hearth), *salow* (water use area), *lansang sid ongkob* (cooking and eating area) and *tingkang* (raised sleeping area).

The open gallery is composed of three sections:

1. Immediately outside the compartment is an aisle (*lansang sid apad*) used by all the members of the longhouse to pass from the entry ladders, located at either end of the longhouse. Though used communally, this aisle consists of sections built, owned and maintained by each family. The aisle is also used for cutting firewood and pounding rice.
2. There is a *salow* along the open gallery. Like the *salow* in the room, it is a wet area. Here, people wash their hands before eating and sharpen their knives (*dangol*).
3. Almost all the daily activities take place at the *tingkang* in the open gallery. Here, women prepare food, weave, tend to the babies, and rock the children to sleep before putting them into the *ongkob* for the night. They also gather here for social functions. Men make baskets, tell myths and stories; young girls and boys play their musical instruments and flirt, and all unmarried boys and visiting males sleep in this area.

The *lansang* and *tingkang* in the open gallery are used optimally while entertaining guests. The traditional dance is performed in the *lansang* while the gong beaters add their rhythm to it from the *tingkang*. *Abai*, the area under the roof, is used to store gongs, jars, rice and seldom used tools.

When a son of the family desires to marry, a substantial bride price is provided for him from the accumulated assets of the family. This, as well as the other institutions which lead up to the marriage and eventually lay the foundations of a new family, are justified by the major value premise in Rungus society: That all sexual relations are potentially deleterious for the participants as well as for the rest of the society unless properly entered into through marriage. Because of this value premise, the sexual services of women are highly valued and scarce, and the explicit and acknowledged purpose of the bride price is the purchase of rights to enjoy these services as well as the reproductive services of the woman.

Gender Roles in Rungus Customary Law

Laura Appell's research (1992) on sex role symmetry among the Rungus of Sabah shows the negotiations that go into arranging a marriage. Most parents will break off negotiations if they feel their daughter strongly objects to the young man. But they will try to bring negotiations to a mutually satisfactory conclusion if they know their daughter is attracted to the boy. Sometimes, the parents decide the man their daughter should marry and try to impose their decision on her. However, if she is truly unhappy about the man chosen for her, she can refuse to accept his advances even after the wedding, a situation which if protracted, can lead to a divorce. *Amu tumutun* is a bride's refusal to accept the husband. (She shuns his sexual advances, ignores him in the presence of other people, refuses verbal communication, does not offer him food or betel, and does not follow him to the swidden.) While the woman has the right to refuse, her parents will try to improve relations between the couple. In most cases, the parents do succeed.

The *buru* (bride price or dowry) consists of gongs, brassware, valuable jars, and various pieces of ceramic wares. A very high value is placed on the services of a Rungus woman as a wife, and the bride price also functions as a social contract to ensure a stable relationship and remind the couple that their relationship must be treated like an expensive item. While there are no studies that relate the value of *buru* to the incidence of divorce, the Rungus believe that the duration of the *buru* negotiations is an indication of the stability of the marriage. Thus, during the negotiations, the father of the bride-to-be will tally all the virtues of his daughter in order to increase the bride price as much as possible. These include her virtue, beauty, comeliness, and skills in household tasks, including sewing and weaving, as well as her diligence in swidden work. A significantly higher bride price will be asked for a woman who is learning weaving and the ritual chants necessary to become a *bobolizan*. The longer the negotiations take, the better it is considered.

While equal importance is placed on the roles of the wife and the husband within the marriage, they are not equal. While the wife and husband are said to balance each other, *mitimbang;* the man is considered heavier (*avagat*). A wife will bow to the leadership of her husband in family matters, during village moots sit alongside him and advise and consent on matters concerning their family. In turn, a husband will follow his wife's advice on religious matters which affect the health and fertility of the family and its domestic economy (ibid.).

The sex symmetry of the spirits in the Rungus belief systems may also reflect the Rungus perception of gender. The basic premise in spirit belief among the Rungus is that the spirits, like humans, have a family or at least work in couples. The *Minamangun* (Creator) when he set off to create the world (*mamangun* literally "to create the world and all living things in it") first had to be prepared for the task by his wife. She wove him a ceremonial shirt and headcloth. The woven design in the shirt included a man, a crocodile and other animals, while the design of the headcloth included lightning, thunder, floods, wind, etc. Until he put on the shirt and the headcloth, he could not create the world and all living things, which he did with a flick of his headcloth (ibid.).

The Rungus customary law has been practiced since time immemorial. For the Rungus and other indigenous people, customary law not only functions as social control, it is also a part of their daily lives, and to a large extent, it manifests how society treats its members. Thus, customary law provides a good window to see the "inner" side of the Rungus society.

One example of the egalitarian treatment of women is the rights and ownership of trees. The Rungus plant trees and bushes, including fruit trees, betel nut trees, and bushes from which rattan may be secured. These trees and bushes can be planted anywhere in the village reserve, but are usually planted near the settlement site, around the edges of the sacred groves, or in the cemetery, so that they are not vulnerable to uncontrolled swidden fires. A person who fells a tree on a grave can be accused of "murdering" the dead and showing disrespect to living members of the family. The person concerned is liable to pay heavy penalties, about half the penalty for murder. However, the planting of a tree does not give rights over surrounding land, and anyone can cultivate the land around the trees.

In terms of intergenerational transfer, the original planter may divide her/his trees among the offspring to avoid disputes after her/his death. However, fruit and other cultivated trees may be devolved upon all heirs, both female and male without divisions. In the case of shared ownership among siblings, there are two kinds of rights to cultivated trees. First, there is the right to cultivate which entails the right to prior enjoyment of the fruit. Then, there is the secondary right held by noncultivating owners to enjoy surplus fruit. The person who holds the rights to cultivate is usually the person who cares for the tree, clears the undergrowth and even builds a fence to indicate the tree is not wild. Thus, the person who has the right to cultivate is usually the person who resides closest to the trees (Appell, 1978).

Roles of Women and Men in the Family

The Rungus believe that the spirit of *Gurau manuk* and *Ponizugung* (spirit of family affairs) determines the marriage of every couple. One of the major criteria considered by the spirit is that the wife and husband should *mitimbang*—"balance each other." Every person has a *timbang* whether or not she or he is spoken for a marriage. The symmetry of roles and their balance is also symbolized by the fact that there is only one term to refer to both wife and husband, *savo;* one term to refer to both mother- and father-in-law, *ivanon;* and *angu* to refer to both sister- or brother-in-law. Similarly, widows and widowers are both referred to as *bituanon.* However, going further away from the circle of marriage, the terms used to refer to relatives indicate their sex. *Odu* and *aki* are used to address grandmother and grandfather, *inai* and *amai* for aunt and uncle respectively.

Another example of sexual symmetry among the Rungus is found in the role of *tandon dot ongkob*. Upon the dissolution of the household of aging parents, they, or the surviving parent, will move in with one of their married children. Ideally, the youngest child (supposedly the last to marry), regardless of sex, would become the *tandon dot ongkob* (see also Appell, 1992). A *bituanon* father may be reluctant to move in with a daughter because during his terminal illness he will be uncomfortable being bathed and cared for by a daughter. Therefore, he may move in with his youngest son. On the other hand, a mother feels her youngest child still needs her help in maturing and usually moves in with that child, regardless of whether it is a daughter or a son.

It can be concluded that the roles of women and men are primarily balanced in the household economy. For each skill exhibited by men, there is an equally important one possessed by the women (see Table 8.1).

Most women and men respondents list cooking, carrying water, tending children and washing as tasks that men defined as "women's duty," while house repair and firewood collection are stated as men's tasks. To my surprise, three men stressed that when men say that the task concerned is a woman's duty, it also indicates that the couple's

Table 8.1
Activities of Rungus Women and Men

Activity	Women's Roles	Men's Roles
Agriculture	• Clearing up small debris prior to planting • Planting • Weeding • Harvesting • *Planting vegetables in their backyard for (mostly) own consumption* • *Handicraft making has reduced women's agricultural involvement*	• Clearing and burning swiddens • Planting • Weeding • Harvesting
Care of domestic animals	• Pigs • Chicken	• Dogs • Water buffaloes
Hunting and gathering	• Gathering snails and shellfish • Fishing with scoops (*sisizud*) for small fish and prawns • Gathering wild roots, nuts and vegetables • *Gathering material for handicrafts*	• Hunting large game with spears • Fishing with traps and nets for large fish • Gathering honey and orchard fruits • *Gathering material for handicrafts*
Domestic Labor	• Husking the family rice supplies • Carrying water • Tending children • Weaving, dyeing, sewing, making rice winnowing baskets, carrying baskets and variety of baskets for general household use	• Collecting firewood • Tending children • Making knives, rope, fish trap, carrying baskets
Property accumulation	• Weaving and selling ceremonial clothes • Payments to the *bobolizan* for curing illness and righting ritual imbalance • *Helping in garden management*	• Marketing of agricultural surplus • Purchase of brassware and gongs • *Managing commercial crops, especially coconut gardens*
Birthing and child rearing	• Ritual aspects of birth • Primary roles in child rearing and nurture	• Midwifery • Secondary roles in child rearing and nurture

(Continued)

Table 8.1 (Continued)

Activity	Women's Roles	Men's Roles
	• *Primary responsibility for child's medical treatment* • *Supporting role in child's religious education*	• *Primary responsibility for child's educational expenses* • *Determining child's religion*
Ritual	(The *bobolizan*) • Communicating with the spirit world through spirit familiars • Ceremonies for health and illness of families • Ceremonies for village	• Ceremonies for swiddens • Ceremonies for property
Christianity	• *Secondary roles in church administration* • *Primary role in Sunday school* • *Generally, more active in church activities as compared to men*	• *Priesthood and priest assistant*
Political activities	• Advise husbands in village moot • *Participate in women's movement*	• Participate in village discussion • Village Head
Recreation	• Social gathering • *Enjoying Hindi movies* • *Sports activities*	• Consume local wine • Traditional sports • Social gathering • *Fishing* • *Watching wrestling (on TV)* • *Modern sports activities*

Source: Modified from Appell (1992).

Note: Activities in italics refer to contemporary Rungus. Most of the time, these are in addition to the traditional activities and not a replacement.

relationship is at its best. In other words, they trust each other, take consensus decisions, and most important of all, respect each other. However, a man could also use this excuse as a polite way to refuse someone's offer.

The *Bobolizan*

Bobolizans are spiritual beings with the ability to heal, work with earth energies and "see" visions. A *bobolizan* is a medium between the spiritual and human worlds. The essential characteristics of *bobolizans* are mastery over chants and rites to communicate with supernatural powers. The ability to memorize every part of the ritual will determine the level of her knowledge and her ability. Basically, the *bobolizan*'s main function is to maintain law and order in the supernatural world. Therefore, her responsibilities are: (*a*) to prevent any behavior which can trigger the spirits' anger by constantly reminding villagers to watch their behavior; (*b*) to restore harmony between humans and spirits by offering animals (blood); and (*c*) to maintain the harmonious relationship between humans and spirits by conducting rituals for every important occasion. However, while the *bobolizan* is a spirit medium, she is not a mediator, because her role is merely to ask for permission, guidance, explanation, authority and "forgiveness" from the good and, most of the time, bad spirits.

There are many versions of the origin of *bobolizan* and *rinait* (the ritual chants used to communicate with supernatural powers). Every village (in this study) has its own version. It is, however, generally accepted that women learnt the *rinait* from a fish. Some versions state that the fish directly taught the women, while other versions suggest that women secretly learnt the *rinait* when they overheard a school of fish teach each other in the river. What is important to note is that men make no contribution to the origin of *rinait* and the *bobolizan*.

In traditional Rungus society, the *bobolizan* may, at times, fulfil the role of priest, metaphysician or healer. The distinguishing characteristic of the *bobolizan* is an ecstatic trance state when the soul of the *bobolizan* leaves the body to ascend to the sky (heavens) or descend into the earth (underworld). The *bobolizan* uses spirit helpers (the *lumaag*), with whom she communicates, while retaining control over her own consciousness. The ability to consciously move beyond the physical body is the specialty of the traditional *bobolizan*. These journeys of the soul may take the *bobolizan* into the nether realms, to higher levels of existence, or to parallel physical worlds, or to other regions of this world. *Bobolizan*'s flights, in most instances, are not an experience of an inner landscape, but a flight beyond the limitations of the physical body.

While most *bobolizans* in traditional Rungus societies are women, men too may have become *bobolizans*. However, only a woman is able to reach the highest status of *bobolizan*—the *rampahan* (the accomplished one). Traditional *bobolizans* developed techniques for lucid dreaming and what is today called "out-of-body experience."

Knowledge of other realms of being and consciousness and the cosmology of those regions is the basis of the *bobolizan*'s perspective and power. With this knowledge, the *bobolizan* is able to serve as a bridge between the mundane and the other states. Few indeed have the stamina to venture into these realms and endure the hardships and personal crises that have been reported by or observed of many *bobolizans*. The *bobolizan*'s role and status among the Rungus is better understood by looking into the Rungus belief system. The Rungus belief system consists of three elements: the good spirit, the bad spirit, and the human being.

Figure 8.2 shows that the Creator (*Minamangun*) is the highest authority in the spirit world. The *Minamangun*'s primary role is to look after the humans in his creation. The Creator's authority was delivered through the *Mononontog* (spirit of birth and death), the *Lumaag* (spirit for public safety), *Mongolungung* and *Mongintanau* (spirit for economy and prosperity), and *Gurau Manuk* and *Ponizugung* (spirit for family matters). Besides these, there are two other spirits whose role is to deliver warnings and signals to human beings.

Unfortunately, humans also have to deal with evil spirits. Ideally, if humans are able to behave according to the evil spirit's wishes and requirements, everything should be fine. However humans are unable to meet the evil spirit's requirements all the time,due to mistakes and other human factors. Furthermore, the requirements and tempers of individual spirits are varied. Therefore, the relationship between humans and the evil spirits is very unstable. The *bobolizan* is the rescuer who stabilizes this relationship. With help from the good spirits, the *Lumaag, Gurau Manuk* and *Ponizungung* etc., the *bobolizan* becomes a mediator who ensures that the evil spirit's requirements are always fulfilled by humans. At the same time, she informs and performs the good spirits' requirements for individual or group concern. In other words, the *bobolizan* becomes the regulator who makes sure that the peaceful relationship between the powerful evil spirits and human beings is maintained.

For example, *bobolizan* play very important roles during the wedding ceremony. The *bobolizan* must ensure the evil spirits are "fed well" before the groom reaches the bride's house. The *bobolizan* will instruct the bride's family to build a temporary hut for the evil spirits where they will be offered a feast. On the assumption that they have already reached the bride's house, the evil spirits will not come to the bride's real house with the groom. Otherwise the evil spirits will create unwanted incidents during the wedding reception. On the

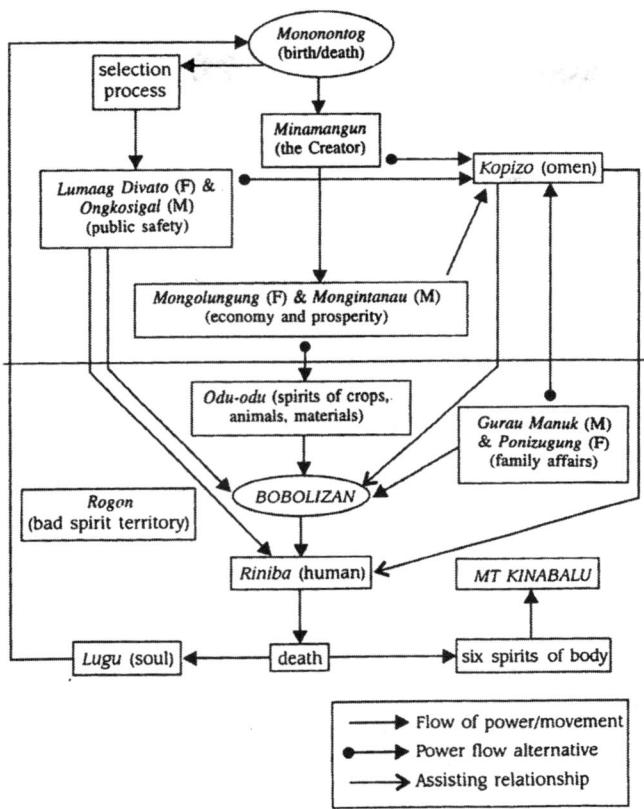

Figure 8.2 The Rungus Spirit World

other hand, *bobolizan* will wait for the groom on the ladder at the bride's house and request the spirits of family affairs to look after his behavior so that the couple will live peacefully and be free from marital problems. The *bobolizan* will perform this ceremony by combing the groom's hair with oil, symbolically to remove and straighten out his unwanted behavior.

Being a *bobolizan* is almost exclusively a woman's task although the term *bobolizan* does not have any gender connotations. It takes many years to memorize the texts that are recited at the ceremonies or rituals. So it was considered more convenient for women or girls (aged about eight or nine years) to learn it as they spend more time at home compared to these men and boys. It would be considered a shame if men, especially unmarried men, were still around after sunrise. Men were expected to do something outside the longhouse, go to their swidden, hunt, collect forest products and so on. In traditional Rungus, the typical son-in-law would be around only after sunset.

While men were structurally denied the opportunity to learn the elaborate ritual texts, there were no social sanctions against a man who is interested in learning the texts (if he can find time after completing his other obligations). Men are, however, unable to reach the accomplished stage because of their inability to enter the trance state and communicate through *lumaag* or by themselves with the spirit world. During the ritual, men helped with the preparations by building a temporary structure, collecting firewood and cooking. The host would choose a handful of unmarried women and men, called *kinambo*, to assist in all the rituals.

In the Rungus community, there are two prerequisites that need to be fulfilled for a woman to qualify as being perfect. One is *rampahan* (expertize in rites), and the second, *boduvan* (expertize in weaving, particularly the *surip* motive). Women who attain these skills are recognized by the community. This is shown by the more expensive *buru* paid for them. A man whose wife is a *bobolizan* is considered very lucky because the woman has a stable income, and can support the family with the *suul* (payment for service) received. The community's recognition of the important role played by the *bobolizan* also makes the man proud of his partner. Most of the *bobolizan* are married. Although there are unmarried *bobolizan*, the married ones are considered better, as they are more mature and experienced.

According to a respondent, the profession of a *bobolizan* is akin to a community nurse, she is often summoned to carry out various rituals. These rituals are usually performed in her village, but other villages also often call for her. The period when the *bobolizan*s are out of the house depends on the type of ritual, but usually if they are summoned out of their village, the period is not less than three days. The main principle in the *momurinait* ritual is that it has to be continuous. If it is a seven-day ritual, then the *bobolizan* has to *momurinait* day and night without stopping. It is also common for a single *bobolizan* to *momurinait* alone for this period of time. When a *bobolizan* is called out from her village, this shows a high recognition for the *bobolizan* and also of the fact that the ritual is of utmost importance. In cases where the *bobolizan* has to leave her house, the husband will take over her responsibilities in managing the household, especially if there are small children. If the children are grown up, the responsibility is given to the eldest child. If her children are breastfeeding, the *bobolizan* will take the child along. A *bobolizan* will carry out certain rituals with the assistance of other *bobolizans.* Therefore, if she is exhausted she would be able to rest (or breastfeed her child). If the *bobolizan* has no children, the husband can usually join her as an honored guest. In some cases, the *bobolizan* can also leave her children with her parents, in-laws or neighbors in the longhouse. If the husband cannot join her, he can come later to assist her in carrying the *suul.* If the husband is unable to come at all, the host will request his *kinambo* (assistants) to deliver the *suul* and his duty is complete only when the *bobolizan* reaches her house.

It can be said that the understanding between the *bobolizan* and her husband is strengthened by the community's acceptance of the *bobolizan's* role in society. A man married to a *bobolizan* is respected by the community. However, there is no rule or *adat* that has to be adhered to by the man who marries a *bobolizan.* Generally, the husband is well informed about the *bobolizan's* role in the community. If the husband has any reservations or is envious of his wife, then the *bobolizan's* opportunity to improve her knowledge and experience will be lessened and the public will think twice before inviting her. But these cases seldom occur. Usually, the husband encourages his wife to fulfil her social obligations as well as collect *suul* which is considered very important in meeting their family needs. There is even an *adat* which enables the *sanganu* (the person who initiates the ritual) to accuse the *bobolizan* if death occurs, and it can be proved that it is the result of the *bobolizan's* reluctance to carry out a ritual. The punishment is half of the punishment for murder.

As an influential figure in the society, a *bobolizan* also commands great respect from her daughter. It is common that the *bobolizan* becomes the role model for her heirs and other young women in her village. The children are also proud of their *bobolizan* mother because of the payment she receives after performing or conducting rituals. The *bobolizan's* ability to cure sicknesses is also very convenient for her married children. However, no "discounts" are allowed and her children also have to pay her, because if the payment is not made, the spirit will transfer the sickness to the *bobolizan.* However, the *bobolizan* can help her children by secretly giving the payment to them (a weeding knife) the previous day. They will then give her the due payment after the curing ritual. Through this secret arrangement, the *bobolizan* fulfils the payment requirement without taking any real payment for her maternal services.

Commonly, the *bobolizan*'s daughter inherits her mother's job. Of the 15 *bobolizans* interviewed, 13 learnt from their mother, one from her aunt, and one from her grandmother. There are no restrictions, and anyone who wants to become a *bobolizan* can do so. About 80 to 90 percent of the women and men aspiring to be *bobolizans* stop halfway and never reach the highest stage in the *bobolizan* ladder. In fact, most students end up with only the most basic knowledge of *rinait* which is useful for minor purposes. The sign of a good student is the ability to memorize the *rinait* after listening to it only once, and recite it a few days later! There are three levels of *bobolizan* ability: *moginsuhut* (practical); *sumahau* (assistant); and *rampahan* (accomplished). Only the smartest student can be accomplished in *rinait*, given the unsystematic teaching methods.

Students have to give certain items to the *bobolizan* after the learning process is over: *tiningkol* (traditional thread) as a payment to the *bobolizan*; *sinontotok* (brassware) to increase endurance in learning; *gapas* (traditional cotton) to increase memory ability; and *kain siam* (piece of cloth) to cover the eye of the bad spirit. A new student is not even considered to be the lowest ranking *bobolizan*. First, she will learn short and easy *rinait* for daily use. After a few years, she will be promoted to the practical stage. At the practical level, a student can be called a *bobolizan*, she has considerable knowledge of *rinait* and is permitted to participate in rituals as a follower or listener. This will give her an opportunity to listen and observe the ritual. Sometimes, if the senior *bobolizan* trusts her enough, she will ask her to recite the easy portions of the *rinait*. It takes about five to 10 practices in every ritual to enable her to be promoted to the next level as an assistant *bobolizan*.

The assistant level is where her *rinait* is almost complete. At this level, the student has the privilege of assisting the senior *bobolizan* in the most important ritual, the *moginum*. The senior *bobolizan* will perform the ritual on the roof of the longhouse while the assistant *bobolizan* will stand below her on the ladder. Again, this will provide her with an opportunity to observe and learn.[1]

In the childbearing age, a woman will not be promoted to the next level. It is believed that the supernatural power will enter her body permanently and the fetus will die, or worse, she will be unable to bear a child unless the *bobolizan* herself is able to neutralize the supernatural power. Most *bobolizans*, however, prefer the first option of not being promoted to a higher level.

The services of a *bobolizan* are also needed to search for *montorik* (amber, which is valued by the Rungus as a great source of wealth and is used in preparing perfumes) within the bark of the *kapur* (camphor) tree deep in the jungle. Due to considerations of time and distance, the male *bobolizan* who knows the *moguhok*[2] will ask for permission from the *rogon* of the tree before cutting it. While only husbands go in search of the *montorik*, the wife plays a very crucial role and her activities at home can determine the success of her husband. For example, she is not allowed to *mongimpuun* (separate the chaff from paddy), *monurud* (comb her hair), express ill-will towards others, or play with sand, or even carry out activities that will bring her hand in contact with soil such as weeding. Her relations with other men in the village must also be carefully observed. All her marital wrongdoings will have a negative impact on her husband's search for the *montorik*. It is also believed that her husband can see her "life" in the tree bark if she is involved in extramarital affairs! Thus her activities are restricted to weaving, winnowing, *mominaig* (cooking porridge, to prevent *mongimpuun*) and *sumagau* (fetch water). The husband must also observe certain restrictions. He should not urinate without proper permission and respect all places and the surroundings. While not every trip is successful, the actions of the wife and husband are regarded as the most important factors that determine the outcome.

The *Bobolizan* and the Village Head

All the respondents clearly distinguish between the roles and responsibilities of the *bobolizan* and community leaders such as the *osukod* or *vozoon*. The *osukod* heads the administrative affairs and is the mediator of the community. He also looks after the daily activities (e.g., when the paddy should be planted, arranging the borders for the *tagad* [ricefield], etc.). The *bobolizan*, however, heads matters pertaining to the relationship between the community and supernatural powers in the belief system of the Rungus people. However, the *osukod* is also associated with supernatural powers, through his ability to communicate with his *ansamung* (personal spirit who can be called on anytime). Some important rituals such as the death ritual, involving supernatural powers are handled by the *bobolizan*, but conveyed to the community through the *osukod*. Many respondents argued that the *bobolizan* has greater influence than the *osukod*. For instance, the village head's decisions can be questioned, but if the *bobolizan* makes a decision (based on her conversation with her *lumaag*), no one can question her, not even the village head.

The village head's role is limited to interpersonal relationships, i.e., problems are solved through the *mihukum* (village assembly) where the accused defend themselves and appoint a defender who is usually an expert in local customs. The process is only completed when one party is unable to debate and contradict the arguments of the other. When this happens, the victor claims compensation according to the rates specified by custom, while the loser tries to mitigate the claim by presenting other customs and humane considerations such as capabilities, age or burdens until both parties reach an agreement. In short, the village head must have the quality to argue and counterargue according to the customary law. The qualities of common sense and logical thinking will enable him to manipulate the existing customary law to his advantage.

The role of the *bobolizan* however covers more important matters such as the future of newlyweds, fertility of the cosmos, safety of the village from the threat of supernatural powers, and healing. Unlike the village head, the *bobolizan* has absolute power in making a decision that has to be carried out by the community. The *bobolizan* can proclaim whether their dwellings are safe or not, delay a wedding ceremony, or order the performance of a ritual. The *bobolizan* can even prohibit the lighting of fire for a specific ritual or any other activity.

Therefore, a good village head will be among those close to the *bobolizan* or will be knowledgeable in matters of customary affairs which are usually decided by the *bobolizan*.[3] The village head is like the pillars of a house, while the *bobolizan* is the different components—showing the importance of the *bobolizan* as compared with the village head in the community (author's personal interview with a village head). Although the *bobolizan* has a role that is bigger than the headman's, some respondents stressed that the respect given by the public is not based on the "job" status (as a *bobolizan* or village head) but on the quality and personality of the leader. As human beings, they will not be respected if they go against the culture, values, norms, and social ethics.

A man could also be a good *bobolizan*, although inferior to a woman with the same qualities. Likewise, women too can become village heads if they fulfil the necessary requirements and social expectations. Many respondents were able to name women who were village heads in the recent past, like: Onjudan, Korinjang, and Boriyad. Boriyad was said to have two husbands. Korinjang reportedly was an expert in customary law and a very brave woman. Another woman, Ungkasul, was not the village head, but was very skilful in hunting, and better than most of the men in her community. According to Sinahaya, a *bobolizan* in Panudahan, even though women could become the village head if they are capable, the men have three advantages: *kipanau* (men are more mobile, stronger, and better equipped to spend the night outside), *kihukum* (men possess more skills in customary matters, compared to women who are more skillful in handicrafts, homemaking or *rinait*) and *orongod* (men look better as the leaders).

In general, a woman can still participate in village matters and the *umpug* (meeting), but many respondents believe that men have built a social structure which prohibits women from interfering with men's business (i.e., matters of interpersonal relationships and village security). Generally, women's role in the meeting is limited to matters pertaining to belief, which will not be questioned by the men.

Usually, as there is only one *osukod* in a particular village, he has no competition. Occasionally an *osukod* also has influence in neighboring villages if he has special abilities such as being a competent mediator and conflict solver. The *osukod,* in making his decision, will be assisted by the elders (including the *bobolizan*). However, the number of *bobolizan* in a particular village is not limited. Ideally, each village should have one *bobolizan*. If it does not have one, the village has to call a *bobolizan* from another village to perform its various rituals, especially the *apansol* (big) rituals. Therefore, a village with many *bobolizans* will improve its image from the religious perspective. This also gives the community a wider choice of *bobolizan* to choose from in either aiding them, or seeking apprenticeship in becoming a *bobolizan.* There are also disadvantages in having too many *bobolizans* in a village. While rivalry among the *bobolizan* is regarded as improper and unethical, it does happen. But it must be stressed that a male *bobolizan* has no place in this competition. It is a woman's affair. The senior *bobolizan* arouse jealousy, especially among newly trained *bobolizans*.

This competition occasionally causes the desperate *bobolizan* (who has no customers) to involve herself in rituals in her village although she is not called upon to do so. If this happens, the *sanganu* usually will not reject her and will politely acknowledge her presence as if she is invited. In fact, this uninvited *bobolizan* will also be given the same amount of *suul* as the invited *bobolizan,* if the *sanganu* can afford it. One *bobolizan* explained that only the invited *bobolizan* should receive the *suul,* but now some *bobolizan's* insist that they should be given *suul* even if they are not invited. Given the respect the community has for the *bobolizan,* they rarely question her.

The uninvited *bobolizan* is, however, deterred by a general consensus amongst the *bobolizans* that the *rinait* will be more complete with fewer *bobolizans.* If there are too many *bobolizans* involved in a ritual, they would rely on each other and parts of the *rinait* may be left out, particularly when it is long and complicated. Usually, the villagers will choose the *bobolizan* who shows more concern for people's economic capability. An understanding *bobolizan* will try to minimize the *suul* or make the payment herself by admitting to the *rogon* that she received the *suul* from the *sanganu.* This neutralizes the competition among the *bobolizan.*

The *osukup* is a person who can control supernatural powers to his advantage or has exceptional traditional medical knowledge. However, when the word *osukup* is mentioned, the first thing on the respondents' minds is that the *osukup* is a killer. The community has stereotyped the *osukup* as a person with no place in the society. In one case, the family members (wife and children) of the *osukup* even asked for public assistance in killing the *osukup* as they were unable to bear his wrongdoings. However, a good *osukup* is valued immensely. In fact, if a *bobolizan* is unable to cure a person, her last hope is the *osukup.* This is because a *bobolizan* is only able to solve a problem pertaining to direct relations with supernatural powers, and not problems arising from supernatural powers driven by human interest. The *osukup* possesses certain skills and knowledge of medications, tracing the origins of the disease, immunity, etc. Every *osukup* must possess the *ansamung.* The respect accorded to an *osukup* depends on his personality.

Hunting the *Bobolizan*

Despite the fact that the *bobolizan* is powerful and is a resource manager, the *bobolizan* herself was also hunted by men to provide a resource for the community. The *bobolizan* was the target of *mangazou,* a person or a group of people who killed other human beings to demonstrate their manhood, strength and bravery. The *bobolizan's* right collarbone, holding the *sovion* (a *bobolizan's* handheld bell that is shaken during the ritual) is said to be an amulet bringing wealth, immunity etc. It is also said that all the *bobolizan's* power (including her melodious voice) could be acquired by taking her right collarbone (which meant having to kill her first).

Unmarried *bobolizan* were often the victims. Since she was especially at risk while conducting the *moginum* ritual, it became customary for a *bobolizan* to be accompanied by a strong young man to protect and assist her in the ritual. This man was also required to wear the traditional costume. The strongest evidence that the killing of the *bobolizan* did occur is the availability of customary law, which requires a man must accompany (protect) the *bobolizan* during important rituals such as *moginum* (held to feed the spirits) and *mamasi* (to cure a very sick person).

The killer's main objective is to show his manhood through his bravery and daring. To do so, he will kill the *bobolizan* during the ceremony while she is holding a *sovion* (in the middle of the crowd) because this is the moment when the ceremony reaches its climax. It is shameful to the villagers if the *bobolizan* is killed in their own village. Therefore, villagers try to keep the killing of their *bobolizan* a secret although it is quite difficult to do so. It is thus difficult to know the actual number of *bobolizans* killed. The *bobolizan* killer will never become a village headman. While his act demonstrates his strength and bravery, it also indicates that the person is big-headed, selfish, inhuman, and antisocial. However, his action can act as a signal to other villages that he can protect his village. To a certain extent, he has gained respect, but this will not advance his social status further.

Hunting parties may involve more than one person. However, since the prime objective of killing the *bobolizan* is to show his manhood (strength and bravery), a man with an excessive ego, who wants to demonstrate his ability, may chose to do the killing alone. The killer may use the *bobolizan*'s collarbone as an amulet for fortune and other supernatural powers (such as invincible strength). The major portion of the collarbone however is used as proof to his future father-in-law and his future bride that he is capable and willing to do almost everything to become a husband/son-in-law. A bride hoping to become *rampahan*, or her family, often asks her future husband to kill a good *bobolizan* and "steal" her melodious voice and knowledge of *rinait*. The melodious voice attracts spirits during ritual performances.

The Rungus Forest Management

Among the indigenous people of Sabah, nature is a major part of their traditional belief systems. Rituals were a meeting point for formal communication with the nature's spirit. For example, in the *moginum* ritual, the *sanganu* will try to sacrifice as many as 12 pigs for every spirit in the natural surroundings. The most malevolent or feared spirits will be given priority, while the less violent spirit will be asked to share. The most violent ones are the spirit of air, spirit of house, spirit of water, spirit of sea, spirit of soil and spirit of swiddens, while the less violent spirits are the spirit of tree, spirit of hole, spirits in longhouse structure, spirit of forest, and the Moslem spirit (the spirit of conversion to Islam, who is propitiated by the sacrifice of chickens).

Communication with supernatural powers also takes place informally through the spirit of omen, or the *kopizo.* Omens are very significant in every aspect of socioeconomic life, such as selecting a site for new swiddens, a site for a new longhouse, wedding, hunting, travel and so on. For example, the appearance of certain animals during the selection process is seen as an omen, indicating either the suitability or unsuitability of the area for cultivation. The suitability of an area for cultivation is indicated, for example, by an encounter with river turtles (symbolizing endurance and strength to survive in any condition) or deer (indicating prospects for good harvest since *tambang* sounds like *kumambang,* the Rungus word for expand). In contrast, rats (*ikus*) indicate a bad harvest because *ikus* sounds like *kumukus* (to shrink). Mostly, the sounds of the names of animals were associated with either the good or bad effects of the task at hand.

Besides animals, the weather (rain, wind, sunshine), sounds, dreams, human interventions or incidents, and animal behavior are also considered to be omens or *kopizo*, which try to inform human beings of the consequences of their actions. The omens also indicate the level of the potential threat or gain involved in the activities. The expansion of Christianity however, eliminated the mutual relations between humans and nature. If the deer is spotted in this hunting enthusiastic Rungus society today, the warning sign is likely to end up on the dinner plate instead. The best example to demonstrate the change of worldview is the Rungus perception of wet soil. Traditionally, wet soil was considered a no entry zone. It was believed that the spirit occupying the area could cause serious illness or even death if disturbed. Today, the Rungus consider this area to be the best place to dig a well or use as a source of water. Some even ask around to see if a mineral water company is interested in utilizing underground water. But lately, frequent dry seasons have given rise to the idea that water is a limited resource and this has stimulated efforts to limit forest utilization and preserve water sources.

For the traditional Rungus, the forests and water are their partners on earth. If there is any restraint in intervening with nature, it is due to fear of evil spirits which occupy almost everything. The *bobolizan*, however, can ask permission from the spirits to utilize their "home." However, some spirits are very sensitive to human disturbance. Physical signs that show the spirits' presence include wet soil, springs, big or small caves, certain tree species, large rocks, strange looking objects or the sighting of omen (*kopizo*) and dreams. Therefore, if a person wishes to open an area for the first time, she must observe the physical and other signs that signify the spirit's requirements and consult the *bobolizan*. Through this process, some areas have come to be recognized as dangerous to cultivate, while others as not dangerous. However, the person still has to perform rituals at almost every stage of cultivation to ask the spirit's permission, guidance, protection, and better yield. Some areas are so dangerous that a trespasser (or his relatives) could fall sick or even die. Through generations, the locals would avoid this area and called it *puru* (a piece of isolated untouchable pocket land characterized by dense forest). In the pre-Christian Era, such protected areas were to be found almost everywhere. Every village had at least one *puru*.

In contemporary Rungus, land is protected not by spirits, but by the human owner. Land ownership gives the individual the authority to exploit his plot and the fate of the *puru* lies with its human owner. While ownership authenticates human authority, Christianity gives them the confidence that the evil spirit is unable to harm them. As a result, the feared *puru* has been cleared to accommodate the increasing demand for swiddens. While most of the *puru* have disappeared, some people have maintained their *puru* for different reasons. In Tinangol village, for example, if the owner (of the *puru* plot) is pagan, it is easy to conclude that the primary reason is fear of the spirit. However, if the owner is Christian, the primary reason is to protect the springs, which are normally found in every *puru*.

In Tinangol, there is one *puru* called *Minapan* (literally, a large flattened rock surface due to centuries of fast flowing water). The *puru* and the surrounding area were surveyed by the land department, and the *puru* was divided among five owners including women, men, old and young couples. Without any official agreement, they imposed a self-restriction: not to disturb the *puru* even though it is part of their land, simply because they were all aware that the only source of water to the larger part of their land was from the *puru*. For them, a water source was far more important than increasing the area of cultivation. They were right. When the long dry season of El Nino struck in 1998, only three sources of water were available in the village: one from the nearby hill, and the other two from their own *puru*. The *puru* saved the village.

In Bavanggazo village, where the Rungus Longhouse Tourism Project is located, the reason for maintaining the forest is very different. Apart from being a source of water, the forest is considered an integral part of their tourism project. Guided treks through the forest are on offer, and these are plans to upgrade and extend the trails. A two-room chalet has recently been built at the edge of the forest to diversify accommodation and cater to a variety of interests. Tourism provides a market for indigenous handicrafts, and this has encouraged the locals to diversify forest utilization. The *igiw* plants, which were formerly considered useless grass, are sought after today for their white beads, which are very appealing to the tourists. *Rinago* (small basket) making has changed the *lingkong* plant (an edible climbing fern) from a popular dish to an important material for handicrafts. The *tontog* (local drum made of wood) was normally made to a strict standard size—2 feet in height and 8 inches in diameter. To satisfy the tourist demand, they are now made in smaller sizes resulting in smaller trees being cut down.

The Rungus had actually reduced the use of raw forest product for building materials. They prefer zinc for roofing instead of *pagung* (kind of palm) and plywood for walling instead of tree barks. However, tourists want to live in the traditional longhouse, and this has increased the demand for traditional building materials. According to one participant in the Bavanggazo Longhouse Tourism Project, the tourists like it more if the longhouse is made of only forest products.

Use and Ownership as the New Basis of Human–Forest Relations

The animist belief system formed the basis of the human–forest relationship in traditional Rungus society. A change in the belief system and the influence of the colonial government has offered a new basis for the interaction between people and nature. Today, the Rungus are not bothered about the fertility of the cosmos and the *moginum* ritual, which was believed to restore fertility in the cosmos (longevity, more children, better yield, good health in humans, animals, plants, and other living beings) is rarely performed. The Rungus have turned to artificial fertilizers, pesticides and weedicides to increase their agricultural production. Thus, the ability to generate cash is a new prerequisite to deal with their increasingly difficult environment.

For the Rungus, forest cleared for shifting cultivation is not considered "opened." It is just *rilik* (trees that have been cut down). The villagers can slash and burn all over their reserve area without opening their land, even if their land is registered under the new ownership system. The land is "opened" only if the person intends to use the *rilik* for commercial and long-term plantations. Today, shifting cultivation has two purposes: to open up land and to strengthen ownership claims by demonstrating efforts to utilize the land. The latter is even more important if the land concerned has no official document as proof of ownership or is owned by the "finger-pointing method." This was a popular method of ownership when individual ownership was first introduced in place of communal land ownership. Practically, the boundaries are marked by physical landmarks such as rare species of trees, hill ridges, rivers, pathways, large rocks and so on.

The land ownership system, which transferred the land from the community to the individual, gave the individual the authority to decide the future of the land. Armed with official documents, the new ownership system has given more authority to the individual to restrict others from using his land. Fenced land is a common practice, and forest products belong to the individual who owns the forest. Today, even if the owner does not use her/his land (in case of an "unopened" plot), other residents or villagers have to ask her/his permission if they wish to use the area for *rilik* or swidden. Normally, the owner will let other villagers cut the trees if they agree to help him open up the area after the harvest. The increasing population and use of cash has encouraged locals to cultivate land even more rigorously. Long-term plantations especially coconut, rubber, cocoa, local fruits, oil palm and short-term crops such as vegetables, pineapples, sugarcane and banana are common. It is interesting to note that the Rungus preferred coconut trees simply because the land was still available for shifting cultivation.

Private land ownership also has other consequences. Traditionally, the longhouse is one of the major social groupings of the Rungus. The longhouse is a symbol of the solidarity and cohesiveness among its members. The urge to "develop" their land has encouraged people to abandon their longhouse and move closer to their private land. To a certain extent, changes in land ownership have weakened social relations. It is common to see an abandoned longhouse with individual houses scattered around it. The change and development experienced by the Rungus of Sabah has a tremendous impact on almost every aspect of their lives. In particular, change in the belief system and property ownership can be regarded as major forces that determine the way they utilize their surroundings, especially the forest. The traditional perception of the forest as their partner on earth has been replaced by a profit-oriented perception where nature only places limits on human activities and is waiting to be conquered.

End of the *Bobolizan* and Emergence of Male Dominance

Looking back at the changes experienced by the Rungus, there is no doubt that there has been a transition in gender relations from egalitarian to a male-centered system. Traditionally, the belief system controls almost every aspects of Rungus life and the system was anchored by the *bobolizan*, an institution that upheld women's status and authority in the community. The level of her knowledge in *rinait* equals her expertize in weaving; both highly valued by society. Therefore, any woman could excel in institutions highly recognized by society, and if she had to compete, her rival would be another woman. The social recognition of the *bobolizan* institution provided a permanent platform for women to make their presence in decision-making visible. The Rungus belief system shows that even spiritual beings are "couples," symbolizing that women and men are equal, and work side by side.

The arrival of foreign influences marked the beginning of the decline of women's decision-making capacity in society. Influence of colonialism in administrative aspects, especially land ownership, has dented the communal approach in dealing with resources. The influence of monetary economy introduced by Chinese traders has opened up new questions: Who should keep the family's monetary income? In terms of land distribution, the same question arises: To whom should the land title be given? For a society that was too naïve to deal with both new issues, the best way to accommodate the new elements is by accepting the instruction, "advice" or the practice of the foreigner. One can imagine, how a Chinese trader would advise the Rungus on who should keep the money. Furthermore, men dealt with outsiders, especially when major economic decisions were involved, such as, how to get cash income. As a result, men were given priority and authority in land matters as well as money matters. Women, on the other hand, were deprived of authority by the new practices.

Change in the belief system of the traditional Rungus meant change in their traditional approaches to life and their surroundings. As far as the *bobolizan* institution was concerned, a change in belief systems means that the *bobolizan* institution had to go. Women's role was further reduced with the recruitment of young Rungus men for agricultural training. Later, the farm schools were converted to Bible schools. The intake, however, remained dominated by male students. Men's opportunity to enhance their knowledge in agricultural activities, especially those related to cash crops created the new social construct that men control the family resources, particularly income generated from agricultural products.

Notwithstanding the traditional hunting of *bobolizans* mentioned earlier, the expansion of Christian missionaries in the 1950s marked the start of major changes in the Rungus worldview. From the very beginning, the missionaries dealt directly with the headman leaving the *bobolizan* in the dark on the consequences to their traditional religion. The missionaries used three methods to propagate the Gospel. In the first method, they dealt directly with the headman, through the headmen's council. The second method employed was the teaching and training of the local people. Three types of schools were established: primary schools, farm schools, and domestic science schools. Hostels were also built to accommodate students. The third method was the health care dispensaries operated by the missionaries, called the "ministry." Through this method, the Rungus were given free medical and health care.

The methods used by the missionaries directly reduced the influence of the *bobolizan* in the community. One former *bobolizan* herself declared:

I have more freedom by becoming an ordinary woman.... In today's world, where the *bobolizan's* contribution is less recognized, it is not worthwhile to maintain the *bobolizan* institution.

Also, it is impossible to implement the *bobolizan*'s ruling in Christian surroundings. Furthermore, it is not easy to follow the strict rules for becoming a *bobolizan* especially before, during and after rituals. (Inombungan Mojuru, a former *bobolizan* in Kg. Tinangol)

First, the Gospel was systematically introduced to provide a new set of beliefs, while the Christian perspective freed the Rungus from fears of evil spirits and other superstitions. They could slaughter pigs and fowls, and bring winnowing trays into sleeping chambers without traditional anxieties. In the traditional community, pork and chicken were rarely consumed because these animals could only be slaughtered during certain rituals. If an animal is killed without informing the spirits, or without a valid reason, the *rogon* will smell the animal's blood and be angered. Thus, conversion to Christianity eradicates the *bobolizan*'s influence.

When the young Rungus began to attend school, it was the starting point of the disappearance of the rites. The young ones have new alternatives to gain knowledge and experience. Today, even the *bobolizan*'s daughter goes to school, and is puzzled about why her mother performs rituals! The missionaries caring for the sick also appealed to the community. People saw the benefits of not having to sacrifice a large number of pigs and fowls. Medical aid was easily and freely obtained from the mission dispensaries. Many also realized that serious illness like tuberculosis and malaria could not be cured by smearing fowl's blood on the patient's ankle, or lime chalk on the forehead. Again, the *bobolizan*'s role as a healer was taken over by the missionaries and the new health care system.

Men's early exposure to Christian education in Bible schools has however changed the traditional division of responsibility where women (through the *bobolizan*) dealt with religious matters and men dealt with nonreligious issues. Christianity has placed both powers with men. The domination of men in the local church workforce, especially at the pastor level, shows that men have an advantage in Christianity. Women's position in the church was redefined in the early 1980s through the women's movement within the church. Later on, women were guaranteed a place in the church executive committee. However, the role of women in the church never equaled their former responsibilities in religious matters, and women still play a secondary role.

NOTES

1. Certain *rinait* such as the *rinait* for the *moginum* ritual are uttered during the ritual only. They cannot be taught during other hours.
2. *Moguhok* is another way of communicating with the spirits, but it is performed in a shorter period of time, in whispers, and without a *sovion.*
3. The customs between localities/*kampungs* are generally the same. However, there are certain rules or *adat* which differ. According to the *bobolizan*, this is because the similar *adats* were sent by the major supernatural powers, while the different *adats* were sent by the supernatural powers of the area. For instance, the earth *rogon* in different areas will have different desires.

REFERENCES

Appell, G.N. (1966) 'Residence and Ties of Kinship in Cognatic Society: The Rungus Dusun of Sabah, Malaysia,' *Southwestern Journal of Anthropology,* 22(3), The University of New Mexico, Albuquerque, pp. 280–301.

––––––– (1976) 'The Rungus Dusun and Other Dusunic Groups,' in *Asia: Ethnographic Studies,* Section 3: Borneo and Moluccas, compiled by Frank M. Le Bar, Human Relations Area Files Inc, New Haven, Connecticut, pp. 1–25.

––––––– (1978) 'The Rungus Dusun,' in Victor T. King (ed.), *Essays on Borneo Societies,* Hull Monographs on Southeast Asia, No. 7, Oxford University Press for the University of Hull, Oxford; pp. 143–241.

Appell, L. (1992) 'Sex Role Symmetry Among the Rungus of Sabah,' in Vinson H. Sutlive (ed.), *Female and Male in Borneo,* The Borneo Research Council Inc., The College of William and Mary in Virginia, Williamsburg, pp. 1–55.

IX

Khasi Women and Matriliny: Transformations in Gender Relations

TIPLUT NONGBRI

Introduction

Since the turn of this century the matrilineal institution in South India has been subjected to a barrage of reforms. Largely initiated by the educated male elite between the 1910s and 1930s, a series of legislations dealing with the rules of inheritance, succession, marriage and the joint family have irretrievably altered the Nayar matrilineal system (Saradamoni, 1996). Though widely separated in cultural and geographical terms from the Nayars, similar demands have now arisen in the Khasi hills, giving a severe jolt to the foundations of the otherwise staid and close-knit Khasi society.

The Khasi are among the oldest inhabitants of India. Edward Gait, the British historian, believed them to be "remnants" of the ancient Mongolian overflow into the country (1967 [1905]: 311; see also Bareh, 1967: 24). Located in the state of Meghalaya, northeast India, the Khasi are best known for their matrilineal kinship system, political deftness and sophistication. However, unlike the Nayars of

Kerala whose matrilineal institutions have been well documented by early administrators and scholars, including anthropologists and sociologists, the Khasi matrilineal system remained in relative obscurity for a long time. Although reference to the Khasi figured in colonial records since the late 18th century, the first ethnographic account on the people appeared only in the early part of this century (Gurdon, 1975 [1905]). A number of books on the history and culture of the Khasi have since appeared on the scene, but few shed light on the pattern of gender relations in the society.

Perceiving that the age-old practice of bestowing descent and inheritance rights on women is detrimental to the male, a section of Khasi men have actively sought to replace the matrilineal with the patrilineal system. Attempts by the state to codify customary laws and practices through the Khasi Social Custom of Lineage Bill (Government of India, 1997) have given a further impetus to this movement by the pro-changers. Although the final legislation to this effect is yet to be passed, recent trends in the hills send out disturbing signals, especially in matters pertaining to women.

The movement not only attempts to exterminate an age old institution and thereby rob women of the traditional security and support provided by the family, it also seeks to redefine ethnic identity in a manner that is highly detrimental to the interests of women. It is interesting to note the way in which the pro-changers are using ideas of progress and social justice to bolster their demands for reform. While their argument on progress may appear to have some legitimacy given the stagnating conditions of the Khasi economy, this has little to do with matriliny *per se* as with the indigenous mode of production and the Khasis' disadvantaged position within the wider social and ecopolitical structures of the nation state.

That this movement should have been able to obtain support in the midst of countrywide debate on the proposed uniform civil code that is directed at securing the empowerment of women, and the international affirmation that "women's rights are human rights," is a serious matter. This chapter looks at the position of Khasi women in the wake of this movement and tries to identify the factors that propelled it. The issues raised here are discussed against the backdrop of the anthropological debate on matriliny. While the chapter questions the wisdom of tampering with the time-tested social institution, it also strongly critiques standard anthropological theories that project matrilineal descent groups as inherently unstable. However, the chapter does not completely discount the vulnerability of the

matrilineal system. Rather, it takes the discourse beyond the
standard anthropological model which looks at the problem primarily
in terms of structural features, to incorporate gender as an important
analytical category. But first let us take a close look at the anthropological debate on matriliny.

Disintegration versus Resilience

Two approaches broadly mark anthropological analyses of matrilineal
institutions. The first predicts the inevitable demise of matriliny in the
face of modernization, urbanization, and colonialism. This prediction
rests on the notion that matrilineal institutions are more liable to
change than patrilineal ones when confronted with economic differentiation. This theory can be traced to Morgan (1985 [1877]) who
regarded the matrilineal system as less advanced than the patrilineal
system. Although Morgan's ideas, which define kinship systems in
terms of stages of human advancement, have been widely challenged
for lack of conclusive evidence, the framework he employed provides
the theoretical basis for many important studies. One of the earliest
and most influential works in this regard was Friedrich Engels', *The
Origin of the Family, Private Property and the State* (1972 [1884]).
Engels endorsed Morgan's evolutionary theory of the human family
and advanced the view that patriarchy and monogamous marriage are
a consequence of the accumulation of surplus value and the rise of
the state. In Engels' schema, matriliny not only precedes patriliny, it
is also associated with preclass, preliterate, and stateless societies,
that is societies which are essentially based on a subsistence level of
production and community cooperation. It is this view that underlies
the disintegration theory in social anthropology. This explains why
Murdock (1949), in trying to identify conditions conducive to
matriliny, asserts that power, prestige, and property spell doom for the
matrilocal principle, despite his proclaimed rejection of Morgan's and
Engels' thesis. For Murdock, the erosion of matriliny is closely tied up
with patrilocal residence, as this involves a man in lifelong residential
propinquity and social participation with the father's patrilineal kins-
men. In Murdock's opinion this meant the end of the road for matrilineal descent (ibid.: 206–7).

The idea that economic differentiation spells doom for the
matrilineal principle also finds tacit support in Aberle (1972 [1951]).
Using Murdock's *World Ethnographic Atlas* as the basis of his analysis,

he proceeds to establish that matrilineal descent groups are confined to a narrow ecological niche, primarily in economies of horticultural base, and are rarely found in societies which practice plough agriculture or animal husbandry. Suggesting that matrilineal descent is ill-suited to economies of high productivity, he notes: "Matriliny, it must be stressed, is a special adaptation to certain productive conditions, capable of surviving under other—but by no means all other—conditions" (Aberle, 1972: 702). Jack Goody (1956) too apparently shares this perspective. In his study of the Lowilli, he claims that disparities in income weaken the matrilineal principle (see also Douglas, 1969).

But it is Kathleen Gough (1972) who puts the final seal on the disintegration theory. Using data from works of fellow anthropologists she attests that in 16 of the cases studied, patterns of change corroborate Murdock's assertion. While she concedes that there are great variations in the degree of change within and between matrilineal societies, she opines that, given continued exposure to the same kind of processes, the direction and end of change seem to be essentially the same. Basing her argument on the modernization paradigm, Gough attributes the disintegration of matrilineal descent groups to their gradual incorporation into a unitary market system, in which all produced goods, but more particularly land and other natural resources, and human labor itself, become privately owned and potentially marketable commodities.

On the conceptual plane, the intellectual dominance of the disintegration theory finds lucid expression in Schneider's characterization of the matrilineal system, which he believes to be riddled with internal "strains" (1972 [1951]). Drawing inferences from Audrey Richards' (1950) idea of the "matrilineal puzzle,"[1] he lists several strains intrinsic to the system and posits the view that matrilineal descent groups are inherently unstable. Instability arises because in Schneider's opinion the "natural" affection that develops between the father and his children poses a threat to the solidarity of the matrilineal descent group. As a corollary to this, Schneider also sees the emotional bond and intense solidarity between husband and wife as a major source of conflict. To minimize the conflict he maintains that "matrilineal descent requires the institutionalization of special limits on the authority of husbands over wives."

Schneider's formulation is purely theoretical in content, and he arrives at this conclusion by using patrilineal descent groups as a point of contrast. In Schneider's conception, the bond between father and child in matrilineal descent groups tends to be in direct competition with the authority of the mother's brother, whereas in patrilineal descent groups the bond between the mother and child poses no such threat to the father's authority. Rather, the mother-child bond reinforces the latter's authority over the wife, as alienated from her own unit a wife aligns with her husband thereby effectively contributing to the unity of her husband's descent group.

Several scholars have challenged the disintegration theory. Rejecting both Murdock's prediction of the inevitable demise of *matrilineal descent groups and Schneider's assumption that the matrilineal system is inherently unstable, they* cite numerous ethnographic examples to argue their point. In their opinion, matrilineal descent groups, far from being riddled with instability and conflict and thereby doomed to extinction, in fact possess a high degree of resilience and vibrancy, which has helped them to successfully adapt to changing conditions. Important exponents of this viewpoint are Elizabeth Colson (1980), who carried out intensive research in the Plateau Tonga, Karla Poewe (1981), who worked on the Luapula of Zambia, and Leela Dube (1996) whose research was conducted on the Lakshadweep Island. These scholars maintain that stability of the family and the kinship domain is associated with women, not with men who are primarily dictated by economic motive.

Poewe shows that among the Luapula there is a sharp polarization between women and men over the matrilineal ideology, with inheritance as the focal point of disagreement between them. She claims that in general men are governed by values of individuality, restricted access to resources and have personal control over investment of wealth. Women, on the other hand, use their wealth to aid their matrikin, which not only reinforce values of cooperation and interdependence between uterine kin but also serves to give them autonomy from their husbands (Poewe, 1981). In a later study, Poewe demonstrates how this differential interest between the sexes also orients them differently towards Christianity. Men look at matrilineal obligations as a drain on their limited resources, hence to protect their commercial interests they prefer to align with Christian religious groups. Whereas women, because of their weak marital bond, see the advantage in retaining their alliance with their uterine kin rather than in the ties established through the Church (Poewe, 1989).

Elizabeth Colson (1980) finds a similar opposition among the Plateau Tonga, which cuts across gender and economic divisions in the society. This opposition placed wealthy farmers and their wives on one side, and small farmers and single mothers who had little to accumulate on the other. Wealthy farmers, who gained from the labor of their wives and sons, fulminated against matrilineal inheritance as it excluded their children from the benefit of the household property in favor of the matrilineal kin. The farmers' wives were also opposed to matrilineal inheritance as it deprived them and their children of the wealth they had worked to create. Small farmers, on the other hand, were opposed to changes in the system of inheritance. As they operated on a limited scale and the family unit consumed most of what it produced, they saw no tangible benefit in switching over to the patrilineal form of inheritance. The same attitude prevailed among the unwed mothers who lived with their parents or matrilineal kin until the time they found a husband; many of them ended up as second or third wives of polygynists. The latter resisted change in the matrilineal system because, lacking in conjugal security, matriliny gave them a claim on the assistance and estates of their brothers and mother's brothers. Colson also found that there was general ambivalence among women and men to the call of the colonial administration for a change in their legal system, to permit either the right of the children to inherit or the right of a woman or a man to dispose of property. Colson sees this resistance to change as an important index of the resilience of matrilineality. She suggests that even among the successful farmers the preference for patrilineal inheritance has less to do with the desire to transmit their property rights to their offspring as with their keenness to possess docile wives and children, who would subordinate themselves to their interests.

Leela Dube (1996) also questions the contention that matrilineal descent groups are characterized by conflict and contradiction, and are therefore vulnerable to political and economic changes. Citing the example of the Lakshadweep islands, she asserts that this Islamic society, which comprises matrilineal descendants from the coast of Kerala, provide an instance of the resilience of matrilineality and its capacity to adapt to a religion with pronounced patrilineal emphasis. In a sharp critique of Richards' idea of the "matrilineal puzzle" and its underlying assumptions, she declared that conflict is not a quality that is unique to matriliny alone, but is a property common to all social systems. In support of her argument, she states that both kinds of unilineal kinship systems have conflict inherent in them, and pointed out that rivalries among patrikins are rampant in patrilineal systems, which become particularly intense when close [male] agnates dispute over property. This betrays the fact that patrilineal systems do not function as smoothly as they are made out to when contrasted with matrilineal systems.

Significantly, Dube notes: "A patrilineal system too is beset with conflict and contradiction in as much as the biologically unrepudiable parent has nothing to do with the group placement of her children" (ibid.: 178). According to Dube, women's peripheral membership in their natal group, their purely instrumental value as bearers of children in their affinal group, and lack of rights over property not only make them vulnerable to oppressions of different kinds, but also give rise to double standards of behavior for female and male, daughter and son, sister and brother, wife and husband, insider and outsider, and the internalization of roles and ideology that circumscribe and devalue women. This contrasts sharply with situations on the Lakshadweep island where, "...women, particularly those with respected kinship statuses such as mother's mother, elder sister and mother's sister had considerable importance and influence" (ibid.: 177–78). Dube attributes this difference to the matrilineal principle, which not only gives women rights over children, property and space, but also greater control over their sexuality. Generalizing from this she opines,

> It seems to be true that women in matrilineal societies are free of any intensive and oppressive control. The relationship between the wife and husband is, typically, not characterized by the authority and control of the man and the corresponding deference of the woman. Asymmetry and oppression are absent from it (ibid.: 177–78).

The two perspectives highlighted in the foregoing discussion appear to be looking at the same problem from two different sides. Both perspectives are concerned with the dynamics in matrilineal societies, but while the first focuses on the weakness of the system, the second focuses on its inner resource and strength. I would argue that the two perspectives are each partly correct in their own way. What the disintegration theory shows is that matrilineal systems are not insular entities, but are part of the wider global structures and processes, their internal cohesion thereby threatened in important ways when forces of change work on the system. The resilience theory, on the other hand, while it recognizes that matrilineal systems are not immutable to change, rightly shows that they are not passive recipients but have a vibrancy of their own to adapt to new situations. This point has been brought home powerfully by Leela Dube in her paper on the Lakshadweep islands.

As analytical paradigms, the disintegration and resilience models are not mutually exclusive; both can be judiciously used to understand the dynamics of matrilineal descent groups in contemporary society. But what is dangerous is when each perspective dogmatically sticks to its side of the story.

Viewed thus, Dube's paper needs to be read with caution. While she rightly explodes the myth, implicit in standard matrilineal theories, that matrilineal systems are riddled with conflict whereas patrilineal systems are not, her preoccupation with resilience and her impulse to generalize lead her to project the lack of gender inequality in Lakshadweep ethnography as a general feature of matrilineal descent groups. Such sweeping generalizations ignore cross-cultural evidence of the increasing devaluation of the system as an aspect of male domination. The drastic transformation of matriliny in Kerala through legislative reforms at the behest of a predominantly western educated male elite (Saradamoni, 1996), and current demands in the Khasi hills to change the kinship system from matrilineal to patrilineal are important instances of this.[2]

That gender stratification cuts across family and kinship structures—though not unmediated by caste, class, race, ethnicity, and age—is clearly established (Palriwal, 1994; Whitehead, 1981; Yanagisako and Collier, 1987). Evidence suggests that even in matrilineal societies, while descent through the mother gives women a fairly high degree of autonomy and influence, it does not guarantee their liberation from gender subordination. The fact that matriliny privileges women in the organization of the family does not preclude them from gender discrimination. This fact is firmly attested by the widespread controversy generated by the recent passing of the Khasi Social Custom of Lineage Bill. The Bill is of immense significance for it not only signifies the turbulent changes gripping the society within which it originates, but it also allows us to take a critical look at the position of women in matrilineal societies. However, before I go into this let me first delineate the salient features of Khasi kinship organization.

The Clan and Family Organization

A distinctive feature of the Khasi kinship organization is the clan (kur), a matrilineally related, exogamous unit, which provides group identity to its members. Membership of a clan is crucial for a Khasi. It not only provides a person with a name (*jait/jaid*) but also confers on her/him rights to citizenship in the society. A person without a clan is considered as *persona non grata* or a social nonentity. As an organizing principle of kinship, the clan regulates marriage, structures the pattern of social relations between kin, and provides the basis of family and religious organization.

For women, particularly, the clan has special significance. As perpetuation of the clan is effected through the female (the Khasi say, *long jaid na ka kynthei*—"from the woman sprang the clan") it is vital that women get married, produce children, and incorporate them into the clan to ensure its continuity. Failure to produce children is a matter of serious concern not only to the woman in question but to the whole family, so much so that infertility and sterility are valid grounds for divorce.

Below the clan are smaller divisions known as the *kpoh* and the *ing*. The *kpoh* (literally womb) is an extended kin group which comprises a group of households whose members trace descent from a single great grandmother. Traditionally, members of a *kpoh* are bound together by shared sentiments of genealogical connectedness, joint land ownership, and a shared cromlech where the bones of the dead are deposited. In the present times, however, the corporate character of the *kpoh* is rather limited. Demographic expansion, spatial mobility, conversion to Christianity and the ultimogeniture pattern of inheritance continually threaten its unity, rendering it functionally ineffective outside the core group of the family.

On the micro plane, the *ing* emerges as the most significant unit in Khasi kinship. Etymologically, the word *ing* refers both to the dwelling unit and the members of the immediate family who are descended from a common ancestress. According to this conception the *ing* is not only a residential unit within which the mundane tasks of production, consumption, and reproduction are effected, it is also the center of family rituals from which non-kin members (i.e., non-consanguines) are excluded. Focused around the mother, the *ing* is premised on the principle of the ritual unity of the sibling group and rests on the cooperation between the youngest daughter/sister (*ka khadduh*) who inherits the property, and the elder brother who exercises control and authority over the affairs of the *ing*. The in-marrying affine, or more specifically the sister's husband, occupies a peripheral position in the ritual life of the *ing*, as ritually he belongs to his own natal *ing* along with his sister and her children. By associating religious function with the *ing*, Khasi matriliny not only lays special emphasis on matrilineal solidarity, it also ensures that brothers have a permanent place in their natal *ing*.

Gender and Matriliny

Many observers have described the Khasi society as matriarchal, implying that women are socially and politically dominant. Charles Lyall, in his introduction to Gurdon's book, attributed the low growth rate of the Khasi to its "matriarchal" system, suggesting that the independence of the wife and the facilities which exist for divorce lead to restrictions upon child bearing, and thus keep the population stationary (Gurdon, [1975] 1905). It is not difficult to understand the reason why Lyall, and many others thereafter, arrived at this conclusion. Given the cultural importance shown to women's reproductive role, their rights over property and domestic space, the relative freedom they enjoy in mixing with the opposite sex before marriage, absence of arranged marriage, high divorce and remarriage rates, and their active involvement in the production process, Khasi society defies standard sociological criteria such as the invisibility thesis or women's oppression, which are generally used to define female subordination.

It would be interesting to ask a Khasi whether the fact that women are central to the family and clan and have rights over property mean that they are matriarchal or conferred with superior powers. A close examination of the cultural notion of women and men, and the division of gender roles shows that the characterization of the Khasi as matriarchal is misleading. On the contrary, there are strong social pointers which show that women are considered inferior to men. In Khasi conception, men are believed to be physiologically and intellectually superior to women. In keeping with this line of thought the Khasi not only invest jural powers in men, but also treat them as protectors of women. The Khasi term *u rangbah* (adult man) neatly captures the high esteem that the society has of men. The term is made up of a combination of two words, *rang* which is an abbreviation of *shynrang* meaning man, and *bah* which means bear or carry on the back. Joined together, the word projects man as a powerful being infused with high moral integrity and the physical and mental ability to shoulder the multiple responsibilities of society.

This stands in sharp contrast to the conceptualization of women. Women are seen to be physiologically and mentally weaker than men, hence they require the latter's guidance and protection. As the "weaker sex," women are not only expected to submit themselves to the control of their brother and mother's brother on the one hand, and the protection of the father and husband on the other, they are also excluded from important areas of decision-making both within the family and in society at large.

It is interesting to note how the Khasi employ the cognitive category of language to establish male supremacy. Anyone familiar with the Khasi language will be struck at the number of gendered terms and phrases in its vocabulary, which symbolize the unequal power relations between the female and the male. To express the difference in physical and mental attributes between women and men the Khasi say, *u rangbah khadar bor* and *ka kynthei shibor*, which means, a man has 12 units of power/energy whereas a woman has only one unit. Further, to reinforce their physiological and intellectual inferiority women are analogically equated with children. In an oft-quoted Khasi simile, *ka kynthei, ka khynnah* (the woman, the child), women and children are clubbed in a single category.

By using the agency of language to legitimate the structure of hierarchy, men ensured a firm ideological base to subjugate and control women. This fact is clearly borne out by the dual structure of authority within the family, the definition of women's rights over property, and their traditional exclusion from the political domain.

The Khasi family occupies a rather unique position in the annals of history for the way in which authority within the household is divided between the dominant males belonging to the two sides of the family. Despite the centrality of women in the organization of the clan (*kur*) and the lineage (*kpoh*), authority over the family does not lie with the mother but is shared between the mother's brother and the father. Although the Khasi vests authority with the mother's brother, the father is not devoid of power. Given that marriage prescribes the coresidence of the conjugal pair, the father not only occupies an important place within the household, as the provider of the family, he also exercises considerable influence over his wife and children. Commenting on the position of a man in the Khasi family Gurdon ([1975] 1905: 78–79) wrote:

> Notwithstanding the existence of the matriarchate, (sic) and the fact that all ancestral property is vested in the mother, it would be a mistake to suppose that the father is a nobody in the Khasi house. It is true that the *kni,* or mother's elder brother, is the

head of the house, but the father is the executive head of the new home, where, after children have been born to him, his wife and children live with him. It is he who faces the dangers of the jungle, and risks his life for wife and children. In his wife's clan he occupies a very high place, he is second to none but *u kni*, the maternal uncle, while in his own family circle a father and husband is nearer to his children and his wife than *u kni*.

Implicit in the above passage is the dual responsibility that rests on the male. As a mother's brother (*u kni*) a man is the jural head of his natal family; he controls the ancestral property and acts as the counselor and guide for his sister and her children in religious and all other matters affecting the household. As a father (*u kpa*), a man is the provider and guardian (*u nong bsa u nong btiah*) of his wife and children. As a social and jural adult (*u rangbah*), a man is expected not only to protect but also to maintain a judicious balance between the interests of his natal and conjugal households.

In ideal terms, this arrangement appears to be advantageous to women, who are the intended beneficiaries of the protection and support of the brother and the husband (father). However, men's dual responsibility not only creates a role conflict for them, it also divides their loyalty between their natal and conjugal families. Also, individuation of interest often leads to a situation where men's dual responsibility deprives the household of the protection and support of both the mother's brother and father at the same time. In such a situation the full responsibility of meeting the needs (social and economic) of the family falls on the woman.

Women's position is, however, somewhat redeemed by the advantage they enjoy in matters of inheritance and residence. In accordance with the matrilineal principle, property traverses strictly from mother to daughter. However, the youngest daughter (*ka khadduh*) alone inherits the ancestral property. Elder daughters may be given some share of the parent's self-acquired property, though this varies from family to family. Among the landed and well-off families, an elder daughter is normally provided with a house plot or basic household goods, but for the rest she has to depend upon her own resources and the provisions made by her husband. The husband's assistance is important as elder daughters usually live independently of their parents.

While the Khasi subscribe to the principle of uxorilocal residence, in practice it is primarily the youngest daughter who conforms to this norm. The ultimogeniture pattern of inheritance allows those without inheritance rights to break away from the parental household to set up independent residence of their own. Though the norm prescribes that the husband joins his wife at her parent's house at the time of marriage, this does not mean that he stays there permanently. If the wife is a nonheiress, the couple remain there only till one or two children are born, or till her younger sister gets married, and then they move out to set up a separate household of their own (see also Gurdon, [1975] 1905).

The youngest daughter and her husband, however, continue to live in the parent's house for the rest of their lives. As the youngest daughter inherits the ancestral property it is vital that she remains in the original home to maintain the continuity of the parental household. However, the fact that the youngest daughter inherits the property neither gives her authority nor makes her the owner of the property. Writing on the nature of the *khadduh's* ownership of her inheritance, Cantlie ([1974] 1934) clearly brings out the limitations of her position:

> *Ka Khadduh* is the custodian of the family property, not the full heir in the sense known in other systems of law, but a limited heir. She is responsible for the performance of religious ceremonies (*ka bat ia ka niam,* she holds the religion); she cremates her mother and if she be the *khadduh* of the whole family, she puts the bones of all members in their final resting place under the stone (*mawbah*) of the clan. The expenses of this ceremony are considerable and, for this reason, she gets a larger share of property or in some cases a piece of family property in addition to and apart from her separate share. Members of the family who are unable to earn for themselves and have no children to earn for them have the right of being fed at the *ing-khadduh.* The actual management is in the hands of her brothers and uncles, and her father is to be consulted. She cannot sell family property without the knowledge and consent of the uncles and brothers. All the sisters have a right to occupy a portion of the family land and *ka khadduh* cannot deprive them of this right (ibid.: 26).

This arrangement imposes severe constraints on the freedom of the youngest daughter who is expected to closely abide by the counsel and guidance of the male matrikin. She not only has to abide by the advice and counsel of the mother's brother who controls the property, *she also has* to fulfill the responsibilities attached to her position, for *if she fails she may be deprived of her inheritance rights.*

Though an increasingly large number of women have begun to assert their independence over their inheritance, in households where a substantial part of the property is handed down as ancestral property, women's independence remains largely circumscribed. As the inheritor of the ancestral property the youngest daughter has to act as the trustee of the family, ensure the execution of rituals, and look after her aged parents and any member of the family who may fall into misfortune or destitution. Above all, youngest daughters are expected to be of high moral standards and act as bearers of their tradition, so much so that poor moral conduct or change of religion could result in loss of their rights. In the past, many women were deprived of their rights due to conversion to Christianity. Today, some groups have even demanded the abolition of the institution of *ka khadduh* as they consider it detrimental to the interest of men.

In contrast to the *khadduh* the elder daughter has more autonomy. Although the authority of the mother's brother extends to all the sisters, the absence of ancestral property in the house of the elder daughter limits his powers over her. But absence of ancestral property invariably increases the dependence of the elder daughter on her husband. Given men's dual position this is not always an easy proposition for the woman. Where the husband is a man of authority in his sister's house, the wife may have to contend with his divided attention between his natal and conjugal households. In her comparative study of the Garo and Khasi matrilineal systems the Japanese anthropologist, Chie Nakane (1967), aptly conveys the piquancy of the situation by quoting an informant who says that he has one foot planted in the sister's house and the other at the wife's. Significantly, Nakane notes the man feels more at home in the sister's house.

Studying the Khasi 40 years after Nakane, and informed by my own experiential location within the system, I would say that though there are wide variations in the way in which men relate to their natal and conjugal households, the basic normative principle remains more or less unchanged. How this affects women and men in their domestic and public lives can be inferred from the high incidence of divorce and single motherhood prevailing in the society, and the lack of concern that divorced men show towards their children. As the children belong to the mother, few men consider it necessary to provide for their maintenance once marriage to the woman is terminated. Ironically, many women hold the institution of *ka khadduh*, which lays the obligation on the youngest daughter to provide support and succour to all members of the family, as primarily responsible for the casual manner in which many men take up their conjugal responsibilities. Men's permanent right in their natal *ing* not only infringes upon their conjugal role, but also provides ample scope for the callous to neglect their parental duties.

The problem is made worse by the dearth of any stipulation in Khasi customary law which enjoins the father to provide for the maintenance of his children. In a recent interview given to *Femina*, Khasi journalist Patricia Mukhim observed:

One of the greatest shortcomings of our state (which unfortunately has a high divorce rate) is the unavailability of free legal aid cells ... so many women abandoned by their husbands have no means of obtaining maintenance for themselves and their children, only because they cannot afford the high legal fees or the long drawn out battles. A free legal aid cell for women is imperative. (Hynniewta, 1998: 132).

Situations also often arise where men's dual role deprives the household of the protection and support of both the mother's brother and father at the same time. In such a situation the full responsibility of meeting the needs (social and economic) of the family falls on the woman. This explains why Khasi women are active players in the production process. In Khasi society women are continually engaged in making and raising resources because in the event of the dissolution of their marriage, or failure of the brother to extend support, it is they who have to provide for the family.

In comparative terms, though, the pattern of residence gives women considerable advantage. Uxorilocal residence provides strong emotional and physical support to women since they do not have to leave their mother's house at marriage for a new and alien environment at their husband's. This is true of elder daughters also, who usually establish a separate household only after spending the initial period of their married life at their parent's house, by which time they are suitably acclimatized to their new status and ready to take up conjugal responsibilities. This practice has prompted some scholars to define the household of the nonheiress as uxori-neolocal because though it conforms to the neolocal pattern, the question of a man taking his wife to his natal house never arises (Sinha, 1971).

Women's comparative advantage on the domestic plane is, however, offset by their exclusion from the political domain. In the past, women could not take part in politics and were strictly excluded from village and state durbars; politics being traditionally held as the prerogative of men. Similarly, women play a subordinate role in the sphere of religion. While women are supposed to make all the necessary preparations for rituals, the actual execution of the rites is done by the men.

One striking role that has particularly been singled out by observers to be the source of women's power and influence is that of the *Syiem Sad*, an exalted office occupied by the mother or sister of the chief (*Syiem*) in the Khasi native states. As the nearest female relative of the chief the *Syiem Sad* is seen as a moral force behind the throne, leading some scholars to define her as the High Priestess and Spiritual Head of the state (Bareh, 1967; Gurdon, [1975] 1905). Since Khasi chiefs are not only administrative and political heads but also custodians of the religious cult of the state, the *Syiem Sad* is seen as the key figure in the rites and ceremonies. Legends about *Ka Syiem* Latympang and *Ka Syiem* Lar, two female *Syiems* who ruled their kingdoms with extraordinary competence in the absence of a male ruler, have also often been cited as evidence of women's power (see in particular Lyngdoh, 1997).

A study of some of the states in the present times has, however, shown that the office of the *Syiem Sad* carries more figurative than functional value. Both in religious and political matters the "powers" of the *Syiem Sad* are subject to a number of checks and constraints. In most of the cases their roles are relegated to the background. For instance, though the *Syiem Sad* is supposed to have a decisive say in the election of a *Syiem,* in reality her consent is a mere formality; the whole process is managed by an electoral college (an all male body) comprising the notables of the state. Indeed, all political and religious offices such as *Lyngdoh, Lyngskor, Myntri, Basan, Sohblei, Sordar, Wahdadar* etc., are held by men. The same is true with the office of the *Syiem.* It is only in very rare and exceptional circumstances that a female is appointed to this office.

Although considerable changes have appeared in Khasi polity and society following India's independence, the political status of women has remained largely unchanged. This is attested by the negligible presence of women in the Legislative Assembly and Autonomous District Councils in the state. Since the creation of the state of Meghalaya in 1972 only two Khasi women (in addition to three Garos) have made it into the State Legislature. The traditional prejudice against women taking part in politics has proved to be a major obstacle to their entry into the modern political process. Women's marginal position in politics has proved highly disadvantageous to their interests. This has not only resulted in the persistence of gender biases in development policies, but more importantly, with the increasing intrusion of the state into the sphere of the family, their absence has allowed men to interfere with their civil and cultural rights. A close examination of the Khasi Social Custom of Lineage Bill and the controversies it has evoked will bring out this fact more clearly.

The Bill and the Ensuing Debate

The Khasi Custom of Lineage Bill, 1997 is the brainchild of the Khasi Hills Autonomous District Council (KHADC). In exercise of the legislative powers conferred by the Constitution under Paragraph 3 of the Sixth Schedule, the KHADC sought to preserve the Khasi matrilineal system by giving it the sanctity of law. Apparently, the act of codification was prompted by the threat of erosion that the Khasi matrilineal system faces in the wake of social change and economic development. Processes of change have not only exposed the Khasi to the values of individualism and personal autonomy, which are opposed to their communitarian and cooperative ethos, but have also resulted in a large influx of outsiders who manipulate the matrilineal system to compete with the local population for their resources.

In a clear reference to this problem, the Bill in its "Statement of Objects and Reasons" categorically states that the legislation is meant "for the protection of their (i.e., Khasi) interests and to prevent claims of Khasi status by unscrupulous persons purely for the benefits, concessions or privileges conferred on the Khasis as the members of the Scheduled Tribe under the Constitution of India." In other words, what the Bill intends to do is to give a legal definition to the word "Khasi." This is deemed to be necessary so that persons who fail to meet the prescribed criteria could be debarred from staking a claim to the Khasi identity, through which they would have been entitled to constitutional benefits meant for Scheduled Tribes.

Twice before (1980 and 1992) the Bill was passed by the District Council, but failed to receive the governor's assent due to memoranda of protests filed by various interest groups and associations.

Among the most critical of these are the Khasi Students Union (KSU), and the Syngkhong Rympei Thymmai (SRT), literally meaning the "Association of New Hearths," who not only rejected the bill but also demanded reform within the Khasi kinship system. Charging that matriliny has a debilitating effect on society, the SRT in particular strongly called for a switch over to the patrilineal system. To bolster their demand the pro-changers made use of ideas reminiscent of Morgan—that the matrilineal system represents an archaic and backward stage of social evolution which should give way to a more progressive system. Interestingly, even scholars like Professor Pakem, the Vice-Chancellor of the North Eastern Hill University, subscribe to this view—a view that has long been discredited by sociologists and historians. Pakem traces the origin of the Khasi matrilineal system to the people's warring past in which men looked after war and politics, and women served as custodians of the family and property. In an apparent show of solidarity with the advocates of reform, he observes,

> We have come a long way from the state of that nature into that of modern economic if not industrialized society. Matriliny is very ideal and well suited for a society of hunters, food gatherers, and shifting cultivators. We are gradually leaving these occupations one after another and preparing ourselves to attain even the status of a post industrial society (in Syiemlieh, n.d.: x.).

What accounts for this disenchantment with the matrilineal system? Or conversely, what is wrong with the Khasi Custom of Lineage Bill that it should set off this storm of protests? Though the discontentment with the matrilineal system had simmered long before the Khasi Custom of Lineage Bill took shape, the two questions are by no means unrelated. Before answering these questions, let us take a quick look at the contents of Section 3.1 which proved to be the major bone of contention and the main factor stalling the Bill.

Section 3 declares:

1. On or after the commencement of this Act, every person born of a legal marriage—

    a. whose parents are or were both Khasi, shall be a Khasi of the *Kur, Jait* or clan of the mother, or
    b. whose mother is or was a Khasi and the father a non-Khasi, shall be a Khasi of the *Kur, Jait* or clan of the Khasi mother, if

and only if the person and his/her Khasi mother fulfil the following requirements, namely:

   i. they can speak Khasi, unless prevented from knowing the language by circumstances beyond their control;

   ii. they observed and are governed by Khasi matrilineal system of lineage, the Khasi law of inheritance and succession and the Khasi laws of consanguinity and kinship;

   iii. had not at any time, in writing or otherwise voluntarily renounced the Khasi status;

   iv. had not adopted the personal law of the non-Khasi father or husband, as the case may be, or a personal law of a society incompatible with Khasi personal laws and customs; and

   v. had not lost or been deprived of Khasi status by judgement or order by the operation of any judgement or order, or under the provision of this Act; or

c. whose father is or was a Khasi and the mother a non-Khasi and the Khasi father and every such person fulfils the requirements specified in Clause (sic) i to v of Clause (b) of sub-section 1 of this Section shall be a Khasi belonging to such Khasi *Kur, Jait* or clan in accordance with the prevailing Khasi customs applicable to such Khasi father or belonging to such new *Kur, Jait* or clan as may be adopted under any prevailing Khasi customs applicable to the Khasi father or by '*Tang Jait*'.

Section 10 of the Bill prescribes that, notwithstanding anything contained in any law in force, a Khasi person shall cease or deem to have ceased to be a Khasi or shall lose or deem to have lost or been deprived of Khasi status, on or from such date as may be specified in the order made in this behalf by the Registration Authority, who voluntarily, in writing or otherwise renounces or has renounced her/his Khasi status, fails to observe the Khasi matrilineal system of lineage or Khasi laws of inheritance and succession, consanguinity and kinship, or has adopted personal law of a society not compatible with Khasi personal laws and customs.

The net effect of this provision is that persons who are considered to have ceased or to have been deprived of Khasi status are no longer entitled to identify themselves as Khasi or use any *Kur, Jait* or clan as their title or surname, or declare themselves as Khasi for any purpose whatsoever.

A perusal of the memoranda of protests, submitted to the chief executive member of KHADC, and debates carried out in private discussions and the media reveal that what led to the widespread controversy is the provision enshrined in the aforementioned 3.1(b), which specifies that children born of a Khasi mother and a non-Khasi father are Khasi.

Technically, there is nothing untoward about this provision as, according to the matrilineal principle, a child belongs to the mother's descent group and never to the father's group. Khasi society has consistently followed this principle of group recruitment since time immemorial, irrespective of the identity or status of the father. As is clear from what has already been said in the preceding section, this does not suggest that the Khasi do not have any concept of paternity, or that they lack rules regulating marriage. Far from it, in fact. As Nakane notes, "The Khasi have numerous prohibitions of marriage between near kin on both paternal as well as maternal sides. The first and second parallel cousins are avoided. Those related to father within three generations are avoided" (1967: 118). Neither is the father an unimportant figure in the Khasi family. While its religious ideology stresses the ritual unity of the sibling group, the household itself is conjugal based. However, for purposes of group placement and kinship identity, it is the mother who matters.

By assigning descent rights to the mother the Khasi give cultural recognition to the woman's role in biological reproduction, which is seen to be more fundamental than the man's role. The Khasi kinship ideology underscores the importance of the fact that the woman not only carries the child\fetus in her womb for ten lunar cycles, but also nourishes it with her blood till it attains fetal maturity. After the child is born she continues to feed it with her milk until it is old enough to subsist on a normal diet. Hence, the Khasi metaphor for common descent is "born of the same womb" and/or "sharing the same blood."

Interestingly, the importance accorded to the woman's role in the act of reproduction confers a special benefit on the non-Khasi women who are married to Khasi men, by enabling them to establish their own clan through which their children are incorporated into Khasi society. As a man cannot pass his descent status to his children, if he marries a non-Khasi woman, his children can adopt their mother's personal name or occupation as their clan name, which in the course of time is recognized as Khasi (see Clause (b) mentioned earlier).

While this arrangement renders the ethnic boundary of the Khasi highly porous, it makes the addition of new members into the society relatively easy and adds to the vibrancy of the system. Unlike patrilineal descent groups where descent and ethnic group membership are transmitted through men only, among the Khasi, while descent is traced through the woman, membership into the ethnic group is transmitted through both women and men. This accounts for the presence of a large number of clans within the society who trace their ancestry to non-Khasi women, many of whom were abducted by Khasi men from the plains in the course of their conquests and trading expeditions, and brought home as wives or concubines.

It is clear that the Khasi descent ideology fully supports the Bill's recognition that any child whose mother is Khasi is a Khasi even though she/he may be fathered by a non-Khasi. So what is the fuss all about? At the root of the problem lies men's fear that recognizing children born of a Khasi mother and a non-Khasi father as Khasi would encourage the proliferation of interethnic marriages, posing grave demographic and economic risks to the society, while also endangering their ethnic identity. In addition to the opposition expressed by the SRT and others through the media, this concern was powerfully voiced by a large section of the participants at a workshop organized by the Hynniewtrep Endeavour Society in December 1997. One group even demanded that women married to outsiders as well as their children forfeit their Khasi identity as well as all associated social, cultural, political, and economic rights in the society.

Spearheading the demand was the Khasi Students' Union (KSU) and the Central Riwar Youth Federation (CRYF) who argued that recognizing such women and their children as Khasi not only exposes the population to economic risks, but also tarnishes their ethnic identity. Underlying the argument is the view that children born of such unions are of impure blood or "half breeds," who undeservedly take up the benefits meant for ethnic Khasi. Hence, in order to protect the interests of the genetically pure Khasi, these children need to be de-recognized as Khasi. In support of their argument some of the participants turned to science. One participant confidently asserted, "Scientifically blood comes from the father and it was due to their ignorance that the Khasi ascribed it to the mother" ("*ba scientifically ka snam dei jong u kpa bad ka wrong assumption jong ki Khasi ka wan rah ia ka jing shim jaid na ka kmie*," HES Report, 1997/98: Panel 1, 4). Needless to say, such arguments not only distort the biological, but also confuse the biological and the cultural.

A close reading of the debate points to the emergence of a new cultural code which informs and redefines the Khasi matrilineal system and the position of women. The demand to forfeit women married to outsiders and their children of the Khasi identity not only departs from the conventionally prescribed mode of reckoning descent, but also reflects a militant strand of androcentric and racial bias. This fact is borne out by the manner in which the demand selectively targets women and leaves out men who are married to non-Khasi women. The KSU and the CRYF, in particular, asserted that children born of a Khasi father and a non-Khasi mother are Khasi, therefore they need not go through the formality of *Tang Jait* to be recognized as Khasi (see Section 3 Clause 1 (c) mentioned earlier), whereas women who marry a non-Khasi should be disqualified from any claim to the Khasi identity, thereby automatically rendering their children outsiders. By privileging the children of a Khasi father and non-Khasi mother, the KSU and the CRYF not only showed their strong patriarchal orientations, but also the primacy of androcentric over ethnic interests.

What seems to escape the attention of many is that, in the name of protecting the ethnic purity of the Khasi and the interests of "pure blood" (Khasi *paka*), they overlook the interests of the castigated women and vulnerable family members who are reduced to the status of outsiders. The only note of caution came from the Meghalaya Women's Alliance, who rightly posed the question, "whether we are ready to reject our own kith and kin when they commit this breach of conduct and deprive them of their rights?" The entire discussion was otherwise dominated by those who view a woman's marriage to a non-Khasi as a slur on the image of the community. An unchallenged assumption that runs through the debate is that women married to outsiders are unpatriotic, devoid of any sentiment or attachment to their culture and tradition. Pointing to the "evils" that marriage between a Khasi woman and a non-Khasi man purportedly brings to the society, the CRYF pointedly asks, "Do they really love the Khasi community after marrying a non-tribal?" (CRYF, Memo 3, 1997).

Implicit in the above question is the message that endogamous marriage, that is marriage between members belonging to the same group, is not only superior to an interethnic marriage but is also expressive of ethnic loyalty. By linking endogamy with loyalty to the ethnic group, this line of thinking not only seeks to secure a firm ideological base for their androcentric policy, it also undermines the legitimacy of the affected to mobilize and fight for their rights. This explains why many women who privately condemned the call to deny children of interethnic marriages of their traditional social rights refrained from airing their dissent in public for fear of being accused of harboring unpatriotic sentiments and values. Thus, caught between the masculinist demand to establish their patriotic credentials and the humane call to protect the interests of their errant sisters and vulnerable family members, they stayed out of the controversy by maintaining a stoic silence on the matter.

The keynote of the debate, however, was struck by Gilbert Shullai, eminent writer and statesman, who in a private conversation with the author observed that the demand to forfeit the ethnic and social rights of women married to outsiders should be weighed against the importance of the clan in the Khasi society. As noted earlier, the clan is not only the basis of their social and political organization, it is also the source of one's identity. To escape from falling into social and physical oblivion a family must be able to perpetuate the existence of the clan, which means that its female members must marry, produce children and recruit them into the matriclan. Failure to produce an heiress by the women of the family leads not only to the extinction of the family (*ing*) but also of the clan (*kur/jait*) which it represents.[3] The cultural importance attached to the clan and the practical exigency that dictates the continuance of the family exerts heavy pressure on women to marry and bear children. So much so that when a woman fails to find a suitor at the appropriate age, family and friends would urge upon her not to be too choosy, or else she would end up an old maid and risk the possibility of discontinuing her family line. Numerous cases show that when the availability of potential mates within the group became restricted, the Khasi are not averse to encouraging their daughters and sisters to accept a non-Khasi as a spouse, provided he fulfilled the social and moral requirements desirable in a life partner.

The Khasi Custom of Lineage Bill appears to have fully anticipated this eventuality, when it prescribes the *condition* under which a child born of a Khasi mother and non-Khasi father can be defined as a Khasi. According to the Bill, a child born of a Khasi mother and non-Khasi father is a Khasi, *provided he/she belongs to the mother's clan and is governed by the Khasi laws of inheritance, succession, consanguinity and kinship,* failing which he/she stands the risk of being deprived of a Khasi identity. The Bill makes it mandatory that all children follow Khasi personal laws, such that if any child adopts the name or customary laws of the non-Khasi parent, the Khasi parent and the rest of the siblings are liable to be forfeited of their status as Khasi. What this implies is that women are basically valued for their reproductive function and ability to perpetuate the family and the clan, therefore their very identity is contingent upon this factor. The Bill not only reinforces this ideology by using the deprivation clause to forfeit the nonconformists of their identity, it also serves as a powerful tool to control and subjugate women.

A close analysis of the Bill and its opposition shows that insofar as their ideology on women is concerned, there appears little that distinguishes the tradition-bound KHADC from the patrilineally oriented advocators of reforms. Sandwiched between the two, women have little choice but to comply with the dominant ideology, which means that they either have to curb their personal freedom and reproduce for the community, or run the risk of being dispossessed of their social, economic, and ethnic rights.

Precursors to the Bill

It would be inadequate to discuss the Khasi Custom of Lineage Bill without looking into the historical and material factors that directly or indirectly contributed to it. The Bill is a sequel to the series of changes that confronted Khasi society since the mid-19th century, all of which have left a deep impact on their culture and tradition. The extension of colonial administration along with its agents, Christianity and education, has not only altered the value of land and land relations, but also sowed the seed of change in the hitherto isolated and economically backward society. While this process has brought Khasi society into the central administrative and ecopolitical system, albeit with some protective mechanisms to safeguard their interests, it has not left the traditional social structure untouched.[4]

Christianity particularly has seriously eroded the ideological and material basis of the Khasi family. While the Christian missionaries did not directly interfere with the customs and tradition of the people, conversion to Christianity threatened the solidarity of the matrilineal descent group. Given that Khasi matriliny is premised on the ritual unity of the sibling group, change of religion weakens the ideological basis of the system and undermines the power of the mother's brother over his natal family. Besides, Christianity with its strong patriarchal orientation binds a man closer to his wife and children thus loosening his tie with the matrikin. This was well illustrated when a Khasi author urged Khasi Christians to be true to their Biblical faith by adopting the patrilineal system (Syiemlieh, n.d.).

It would not be incorrect to say that Christianity not only provided the first institutional challenge to matriliny, but it also reinforced the ideology of female subordination. Despite the values of equality and social justice preached by the church, the involvement of women in its activities are confined primarily to subordinate and stereotype feminine roles, rarely in the ministry. Indicative of their patriarchal orientation, none of the major Christian denominations in the Khasi hills has ordained a woman as a pastor or deacon. Commenting on the marginal position of women in the Christian church, Warjri (1993) ascribes it to: (*a*) the patriarchal cultures and traditions of both Jewish and Greco-Roman societies; (*b*) the androcentric language used by Bible translators and scholars; and (*c*) an androcentric frame of reference. She cites numerous examples from the Bible where women played an important role but were downplayed by Biblical scholars in their translations. These distortions not only helped the church to establish and sustain a monolithic power structure from which women were, as a rule, excluded, but it also helped create a social milieu fostering patriarchal values. As an agent of modernization, Christianity has been able to propagate western patriarchal values even to the non-Christian population through its educational institutions.

The retreat of colonialism, the accession of the Khasi states to the Indian Union and the process of nation-building that followed accelerated the steady erosion of matriliny and the position of women. Attracted by the rich resource base of the region, people from neighboring states came in large numbers to exploit its rich hinterland. While this process intensified interactions between the indigenous and nonindigenous people, it also resulted in a high incidence of interethnic marriages, posing a serious threat to the traditional principle of ethnic group endogamy which regulates and controls women's sexuality.

Breach of endogamy is not only perceived to endanger the demographic balance within the society, in the sense that children born to a Khasi woman from a non-Khasi man may be incorporated into the latter's group. It is also believed to provide effective entry points through which outsiders could gain access to the social and material resources of the community (Syiemlieh, n.d.). Whether there is any truth to this belief needs to be closely investigated. However, what is important is that this has not only put interethnic marriages (primarily of the female variety) under a cloud, it has also brought the women's private life under public scrutiny. The proposed legislation of the Lineage Bill and the demand to deprive women married to outsiders of their social and ethnic rights are directly linked to these processes.

Besides, the process of political modernization activated during the period, the formation of political parties, the movement for a separate state, which led to the reorganization of Assam and the subsequent birth of Meghalaya in 1972, gave rise to new ideological and political thinking and a new awareness about the significance of one's cultural identity.[5] Identity, based on common language, religion, territory, or other shared primordial characteristics not only marks the cultural distinctiveness of a people, it is also a source of emotional and moral security, and a viable tool for political consolidation and support. What this suggests is that identity is not simply a cultural marker but a value to be cherished and preserved both for its subjective and instrumental use.

An important outcome of this new social consciousness was a concerted attempt made by state to conceptualize and define Khasi identity. The task was vital, as this not only determined a person's status, but also her/his right to the social, cultural, and material resources of the society.[6] Seized with the importance of the problem, the Land Reforms Commission, an official body set up by the Government of Meghalaya shortly after the state came into being to examine the land tenure system in the Khasi hills and make recommendations for reforms, worked out the criteria that would qualify a person as a "Khasi." After considering various opinions expressed by early Khasi thinkers on the subject the Commission noted:

> The Commission have given a deep and serious thought to these various views expressed and are of the opinion that it may be conclusively stated that a person who is acceptable as a Khasi in this context is one whose parents descended from time immemorial from the descendants of the people inhabiting Ka Ri Khadar Daloi, Ka Ri Laiphew Syiem, or one who has adopted Khasi socio-political customs and way of life, conducts and comports himself as a Khasi, speaks Khasi language, follows a matrilineal system, and in the case of male adults have a right to take part in traditional durbars of the Khasis in a place where he lives or to take part in the *election of hereditary chiefs of his elaka* where popular election is held *in which women cannot take part*, and is accepted by the rest of the people as belonging to their tribe (Government of India, 1974: 15. The italics are mine).

As a large number of Khasi have embraced Christianity, the Commission was careful to note that those Khasi who became Christians are not debarred from Khasi identity, provided they retain Khasi customs and practices. However, it ruled that the Khasi who had dispersed to other places and had lost their identities may no longer be treated or accepted as Khasi and, as such, had lost all their claims to the privileges and customary rights of a Khasi over property (ibid.: 14). Similarly, Khasi women married to outsiders who no longer follow Khasi customs can no longer claim to own land or inherit property (ibid.: 31).

By privileging culture over biology the Commission made the individual subordinate to the community. Hence, while it recognizes the inevitability of interethnic marriages, it demands that the children born of such unions are incorporated into Khasi society, lead a Khasi way of life, governed by Khasi customs and practices.

This process marked the first intervention by the state into the sphere of the family. For women the process had important consequences, since for the first time the status of children born of a Khasi mother were open to question. It was also for the first time that the concept of illegitimacy gained currency among the Khasi. In Khasi tradition there has never been any dispute over the social and legal status of a child born of a Khasi mother, irrespective of the identity or status of the father. Indeed, A.S. Khongphai, late lawyer and president of the *Seng Khasi*, wrote, "As the children belong to the mother there is no problem of illegitimate children amongst the Khasis" (Synrem, 1992). Reacting to this statement, the Commission observed:

> There is, however, a monstrous problem staring people in their faces—the dishonour, the degradation, and what may turn out to be total extinction of a race with a distinct history, genius and culture of its own. It is therefore a matter for serious consideration if this question should not also be tackled by a duly enacted law under inheritance that illegitimate children are debarred from inheritance (Government of India, 1974: 30).

Interestingly, the Commission predicated its observation on a statement made by David Duncan, an Anglo-Khasi bureaucrat, in his memorandum that, "the race is ruined because of the totally unfettered freedom of women; they are getting fatherless children heedlessly for there is no more control over inheritance" (ibid.: 30).

When we read the Commission's observation along with Duncan's statement three things come to the fore: First, implicit in the statement is the assumption that illegitimate children do not deserve any social consideration. Second, that the fault for illegitimacy lies solely with the woman. Third, that the dishonor and degradation allegedly caused by the woman results in dishonor and shame to the race.

These assumptions are not only colonial in ethos, they are also highly prejudicial to women. Given that conception or the act of biological reproduction requires the cooperation of a male and a female, it is erroneous to lay the blame exclusively on the woman. However, by uncritically accepting Duncan's highly prejudicial statement the Commission not only appears to be grossly indifferent to this fact, it also ignores the vulnerability of the woman and her helpless children. And, by absolving men of any blame in the alleged degradation of the race, the Commission betrays its own androcentric predilection. Although the Commission upheld the status quo in matters of inheritance it would not be incorrect to say that questions raised during the investigation paved the way for the current debate on the matter.

Identity, Sexuality, and the State

An important point which emerges from the previous discussion is the polarization of public opinion into two opposed camps, the supporters of the Bill led by the KHADC, and the pro-reforms championed by the SRT and supported by other independent groups and individuals. But what is important to note is that straddled between the two is the issue of female sexuality, ethnic identity, and the control of property. As woman is the focal point through which descent and inheritance are transmitted it is vital that her sexuality is strictly guarded. Kinship and the community play an important role in this regard. Although the Khasi woman possesses considerable freedom in matters of marriage and divorce, the principles of clan exogamy and village endogamy restrict marital alliances within permissible limits. Further, the institution of the *khadduh*, the powers of the mother's brother to censure, and the operation of strong social codes within what Bourdieu defines as *habitus*, condition individual choices and provide a "feel for the game". These social norms act as a moral force and orient the unconscious drives of sexuality and aggression in a socially acceptable direction through "the hidden persuasion of an implicit pedagogy" (Bourdieu, 1990: 66–67). However, as long as the Khasi remained isolated from the wider population, these forces

operated below the level of daily consciousness, drawing little attention to sexuality or the symbolism of a woman's body, except for the
woman as an embodiment of motherhood.

The advent of colonial rule and the social, cultural, and political
transformations it heralded has pushed the issue of women's
sexuality center stage and made gender identity a politically charged
affair. The woman's body has become an ideological terrain around
which identity and entitlements to property are reconstituted and
reworked.

The controversies surrounding the Khasi Custom of Lineage Bill
are an attestation of this fact and an implicit attempt by men to reinforce their domination in the wake of social change and threat of losing their traditional control over women. To the Khasi, the central
defining feature of their identity is the matrilineal system and the
woman is the bearer of this tradition. Hence, to preserve their identity, it is vital that a woman's sexuality and the products of her body
remain under the control of the community.

However, the subordination of women and/or the control of
women's sexuality cannot be ascribed simply to the matrilineal system. It represents and constitutes part of men's struggle for political
domination, which is fueled both by internal and external factors. As
the aforementioned evidence shows, the process of decolonization
and economic development, the struggle for political autonomy and
rise of ethnic politics among indigenous people in the region have
made women more vulnerable. Perceiving that control of women is
vital to their interests, men have lost no time in coming up with strict
normative codes to regulate their conduct.

Central to this process is the role played by the state in producing
and perpetuating an ideology that subjugates and marginalizes
women. It is interesting to note how in the guise of protecting their
identity and tradition, the state has come up with standards that
grossly devalue women. To the extent that the state machinery is
dominated by men, state policies largely reflect men's interests.
Women's marginal position in the political process has aided men in
this regard. Neither the Land Reforms Commission nor the Khasi
Hills Autonomous District Council had any woman representative
when the matters discussed previously were taken up. Neither did
the Land Reforms Commission appear to have solicited women's
opinion, through their issuance of questionnaires. Out of the 70
memoranda filed by the Land Reforms Commission, only one came
from a woman, who in a petition jointly signed with her daughter,

opposed the government's proposal for land reforms on the ground that it would be detrimental to the interests of the people. Similarly, the Select Committee set up by the KHADC to examine and work out the parameters for the Khasi Custom of Lineage Bill, before placing it before the House, did not include a single woman in its panel.

These facts not only exemplify the highly gendered character of the state, they also point to the selective and partisan manner in which it identifies the issues for legislation. While it is true that there is a high incidence of intermarriage between Khasi women and non-Khasi men, the state makes little attempt to identify the multiplex factors that contributed to the phenomenon. It is an open fact that interethnic marriage is part of the universal process of modernization and urbanization; when people move out of their villages and towns the marriage circle tends to expand, sometimes going beyond the traditional boundary of the village and community.

It is pertinent to note that at the bottom of the question of female sexuality and the valorization of ethnic identity in the current debate, is the role of the state in the degradation of women. Although interethnic marriage occurs in many societies, in the present context it cannot be disassociated from the process of development carried out by the state. State-sponsored development, with its emphasis on building the infrastructure, not only brings along with it a large contingent of labor and other personnel from outside, it also contributes to women's impoverishment by appropriating their land and forest resources, hence exposing them to economic and sexual exploitation by outsiders. Not so long ago, Roy-Burman (1990), noting the active movement of the migrant population in the northeastern states remarked:

> The presence of a large number of sojourners in its turn, coupled with growing imbalances in sex ratio, has serious implication for the quality of life. It means that there is a good number of male population from outside, without strong social moorings in the region. Social workers will have to address themselves to the consequential problem (p. 70).

What Roy-Burman obliquely notes as the consequential problem that requires the attention of social workers apparently refers to the growing number of unwed mothers in the region as a result of the exploitation of indigenous girls by the nonindigenous men. In their struggle for livelihood and security for their family many Khasi women have fallen victim to non-Khasi men, who under the pretext of marriage used them for their own benefit and then dumped them to their fate (Nongbri, 1998). However, instead of booking the culprit and tackling the problem at the socioeconomic and developmental level through raising of consciousness and eradication of female illiteracy, poverty and unemployment which are at the root of the matter, the state has done little in this regard other than impose strictures on the victims.

Also extremely significant by its silence is the state's attitude to persons who indulge in *benami* (fake) transactions, in which nonindigenous businessmen use Khasi women and men as fronts to secure licenses, permits, and other benefits to further their economic interests. It is a well known fact that a large number of persons, many of whom hold important positions in the government, engage in illegal transactions with outsiders, allowing them to escape the arm of the law by lending their names to the latter's business. Though there is a law which prohibits *benami* transactions, no prosecution has ever been made in this regard. It would not be very wrong to say that had the state been sincere in its intention to restrict the misuse of benefits by the non-Khasis, it could have used the anti *benami* law to root out this practice. This would have saved the state from coming up with a controversial bill like the Lineage Bill and also put a check on prospective exploiters of Khasi women. Apparently the fact that it is primarily men who engage in this illegal practice may be responsible for the state's inaction.

This power enjoyed by the decision-makers, to confine public policies and discourse to issues that are integral to their interests, is basically a patriarchal power and corresponds to what Kabeer following Lukes defines as "power over" (1995: 225). By projecting a distorted picture of women the holders of power enable the state to divert attention from its negative practices and push through its own ideological and political agenda. The Khasi Custom of Lineage Bill and the debate that followed reflect a clear attempt by men to use both the state machinery and civil society to perpetuate their control both within and outside the family. While women *married to outsiders* is the issue being contested, the implicit message is that women as a category are the property of the community, hence their sexuality, their children, and their overall deportment and conduct are the concern of men who are its unchallenged protectors and custodians.

Women themselves have little say in the matter; instead they are reduced to being mere objects on which the laws and regulations are to be applied and enforced. Women's subordination is implicated not only by their very absence from the official centers of decision-making, but also by their impotency to resist even at the level of civil society. Interestingly, by projecting the Lineage Bill as a "nationalist" agenda, directed at saving Khasi culture and tradition from the threat of extinction under the influence of the cultural "other", the state is able to stifle potential dissenting voices, who for fear of being branded as antinational and/or unpatriotic, have resigned themselves to the regime of patriarchal control.

Conclusion

This account brings into sharp focus the unequal relationship between women and men among the Khasi. Although women are traditionally disadvantaged, matriliny gave them comparative autonomy and influence and provided them with institutional support and security not available to their counterparts in patrilineal societies. However, matriliny has also been used to add to women's vulnerability. By vesting descent rights on the mother, women are not only objectified as symbols of their culture and tradition, they also become subjected to a strict social and moral code to uphold the honor of the family and society.

This account challenges the idea that matrilineal societies are free of any intensive control. As children are recruited into the descent group through the mother, a woman's sexuality, in particular, becomes the concern not only of the family but also of the community. The process in the Khasi hills clearly demonstrates this fact. Perceiving that women are the entry point through whom outsiders could gain access to their social and material resources, the men have laid strict normative codes to regulate their conduct.

This chapter shows a dense connection between gender, patriarchy and the state, with ethnicity as the mediating element between the three. It is noteworthy that with the process of modernization and economic development and the struggle for autonomy by indigenous people in the region, not only has patriarchal control over women intensified but the relevance of matriliny itself has been questioned. The demand for reforms, voiced by the pro-changers, though couched in the language of ethnicity, is a clear attempt to strengthen and legalize patriarchy. Although overtly the state has taken a contrary position in the matter, as the aforementioned account shows, its policies are neither in keeping with the traditional matrilineal ideology nor with the values of gender equality.

This brings me to the initial theme with which I started the chapter, that is, what implication does the study have on kinship theory? As I had elaborated at the outset, an important concern in anthropology in the last two decades has been the future of matriliny. However, while studies organized around this point throw valuable insights into the dynamics of the system, they shed little light on gender relations. Though more recent studies (Dube, 1996; Saradamoni, 1996) have attempted to rectify this lacuna, they remain largely region specific. Moreover, given that gender is not an abstract entity but firmly grounded within its sociocultural and historical context, gender relations cannot be read off mechanically purely on the basis of kinship rules. This study shows that while kinship provides an important context within which gender ideology is shaped and practiced, it cannot be divorced from wider social structures and processes.

In the final analysis I would say that it is meaningless to talk either about the future, or the current status of the matrilineal institution, without looking into the gender question. As this chapter shows, women are not only the focal point around which the system revolves, they are also the sites upon which women and men's rights are articulated and contested.

However, before I conclude, a note of clarification is warranted by way of explanation for the Khasi woman's apparent lack of agency. This chapter specifically looks at the constitution of gender within the Khasi matrilineal structure and the implications of the Khasi Custom of Lineage Bill, omitting in large part the actual way in which women organize their daily lives. Constraint of space did not permit me to go into this aspect. It would suffice to say that while the responsibility of matrilineal descent enhanced women's agency in the family and the economy, their marginal position in politics restricted them from converting this opportunity to their advantage. It is interesting to note how men used the gender-based division of labor to subjugate women. By projecting the division of labor as a natural aspect of complementarity (for instance, the Khasi say, "war and politics for men, children and property for women") men keep women out of the legal political process through which policies are formulated and implemented. As a result the large majority of women were really unaware about the Bill or the hectic debate it

generated either on gender or ethnic identity. The few who did know
about it were either intimidated or appropriated by the state.

NOTES

1. Richards used this concept to highlight the conflict that occur due to the
   perceived contradiction between matrilineal descent and exogamous marriage.
   In epistemological terms, however, there is a distinct difference between
   Richards' idea of the "matrilineal puzzle" and Schneider's theory of matrilineal
   descent groups. The former used the concept as an analytical category to make
   sense of her ethnography in her study of family structures amongst the Central
   Bantu. The latter on the other hand, seeks to establish broad generalizations on
   the basis of logic and deductive reasoning.
2. Also note the insights offered by Poewe's and Colson's papers.
3. To some extent this eventuality can be mitigated by the practice of *rap ing*
   (adoption), whereby the families so affected can adopt a female uterine kin nearest to the family or an unrelated girl and invest upon her all the rights and
   duties that would have accrued to their own daughter.
4. The protection of indigenous areas constitutes an integral part of the policy of
   the Indian government. Derived from the colonial policy of "exclusion" and "par-
   tial exclusion" enshrined in the Government of India Act, 1935, this policy finds
   expression in the form of the Fifth and Sixth Schedules in the Indian constitution. The Khasi and Jaintia hills come under the purview of the latter. Under the
   provisions of the Sixth Schedule, the Autonomous District Councils within each
   state where they operate, are assigned special legislative and judicial powers
   over indigenous people's land, forests, the appointment of chiefs, the formation
   of village and town councils, and the protection of customary laws and practices.
5. This process is not unique to the Khasi; the whole northeastern region witnessed indigenous people redefining their identity, discarding old names for
   new ones, or entering into new patterns of alignment and realignment with
   each other in a bid to achieve ethnic solidarity and political stability. However,
   ethnicity mobilization may take various forms depending upon the configura-
   tion of cultural, economic, and political factors and interests at a particular
   point in time. In a polyethnic society, where group interests converge, ethnic
   mobilization may take an expansionist form with different groups coming
   together in ethnic alignments and realignments, as among the Nagas. However,
   where interests diverge or where groups compete for scarce resources ethnic
   mobilization becomes reclusive, isolating elements that pose a threat to their
   unity, or that puts a strain on their meager resources, thereby endangering their
   social/political/economic survival. Ethnic and identity formation in the Khasi
   hills falls in the latter category.
6. Two separate election petitions (Wilson Reade versus C.S. Booth, A. Alley and
   Jormanik Syiem; and, A.S. Khongphai versus Stanley D. Nichols Roy) challenging

the candidates' status as a member of the Khasi Scheduled Tribe, on the ground that one of the parents was not a Khasi brought urgency to the matter. (For details see Synrem, 1992: 36–39.)

References

Aberle, D.F. (1972) (originally published by University of California Press, Berkeley, 1951) 'Matrilineal Descent in Cross Cultural Perspective,' in D.M. Schneider and K. Gough (eds.), *Matrilineal Kinship*, A.H. Wheeler and Co., Allahabad, pp. 655–727.

Bareh, H. (1967) *The History and Culture of the Khasi People*, published by the author, Calcutta.

Bourdieu, Pierre (1990) *The Logic of Practice*, Polity Press, Oxford.

Cantlie, K. (1934 [reprint 1974]) *Notes on Khasi Law* (edited by A.S. Khongphai), Ri Khasi Press, Shillong.

Central Riwar Youth Federation (1997) Memo 3, unpublished.

Colson, E. (1980) 'The Resilience of Matrilineality: Gwembe and Plateau Tonga Adaptations,' in L.S. Cordell and S.J. Beckerman (eds.), *The Versatility of Kinship*, Academic Press, New York, pp. 364–66.

Douglas, M. (1969) 'Is Matriliny Doomed in Africa?' in M. Douglas and P.M. Kaberry (eds.), *Man in Africa*, Tavistock, London, pp. 121–35.

Dube, L. (1996) 'Who Gains from Matriliny? Men, Women and Change on an Indian Island,' in Palriwal, Rajni and Carla Risseeuw (eds.), *Shifting Circles of Support: Contextualizing Gender and Kinship in South Asia and Sub-Saharan Africa*, Sage, New Delhi, pp. 157–89.

Engels, F. (1884 [reprint 1972]) *The Origin of the Family, Private Property and the State*, International Publishers, New York.

Gait, E. (1905 [reprint 1967]) *A History of Assam*, Thacker Spink & Co., Calcutta.

Goody, J. (1956) *The Social Organisation of the Lowilli*, H.M.S.O., London.

Gough, K. (1972) (originally published by University of California Press, Berkeley, 1951) 'The Modern Disintegration of Matrilineal Descent Groups,' in D.M. Schneider and K. Gough (eds.), *Matrilineal Kinship*, A.H. Wheeler and Co., Allahabad, pp. 631–52.

Government of India (1974) *Report of the Land Reforms Commission*, Shillong.
——— (1997) *The Khasi Social Custom of Lineage Bill*, Shillong.

Gurdon, P.R.T. (1905 [reprint 1975]) *The Khasis*, Cosmo Publications, Delhi.

HES (Hynniewtrep Endeavour Society) (1997/98) *Ka Jingiaphylliew Jingmut*, Shillong.

Hynniewta, Parnell E. (1998) 'Daughters of the East,' *Femina*, October 1, 1998, pp. 131–32.

Kabeer, N. (1995) *Reversed Realities: Gender Hierarchies in Development Thought*, Kali for Women, New Delhi.

Lyall, C. (1907 [reprint 1975]) 'Introduction,' in P.R.T. Gurdon, *The Khasis*, Cosmo Publications, Delhi, pp. xvi–xxvii.

Lyngdoh, P. (1997) 'The Role of Women in Politics with Special Reference to Meghalaya,' in *Souvenir of the Meghalaya Legislative Assembly*, Silver Jubilee Celebration, Shillong, pp. 78–81.

Morgan, L.H. (1877 [reprint 1985]) *Ancient Society*, University of Arizona Press, Tuscon.

Murdock, G.P. (1957) 'World Ethnographic Sample,' *American Anthropologist*, 59, pp. 664–87.

Murdock, P.G. (1949) *Social Structure*, The Free Press, New York.

Nakane, C. (1967) *Garo and Khasi: A Comparative Study of Matrilineal Systems*, Mouton, Paris.

Nongbri, T. (1998) 'Gender Issues and Tribal Development,' in B. Singh (ed.), *Tribal Self-Management in North-east India, Antiquity to Modernity in Tribal India*: Volume II, Inter-India Publications, New Delhi, pp. 221–43.

Palriwal, R. (1994) *Changing Kinship, Family and Gender Relations in South Asia*: *Processes, Trends, Issues*, Women and Autonomy Centre, University of Leiden, Leiden.

Poewe, K. (1981) *Matrilineal Ideology: Male-Female Dynamics in Luapala, Zambia*, Academic Press, London.

——— (1989) *Religion, Kinship and Economy in Luapala, Zambia*, Edwin Mellen Press, Lewiston, New York.

Richards, A.I. (1950) 'Some Types of Family Structures Amongst the Central Bantu,' in A.R. Radcliffe-Brown and D. Forde (eds.), *African Systems of Kinship and Marriage*, Oxford University Press, London, pp. 205–51.

Roy-Burman, B.K. (1990) 'Tribal Population and Development,' in A. Bose, T. Nongbri and N. Kumar (eds.), *Tribal Demography and Development in North-East India*, B.R. Publishing Corporation, Delhi, pp. 67–83.

Saradamoni, K. (1996) 'Women's Rights and the Decline of Matriliny in Southern India,' in R. Palriwal and C. Risseeuw (eds.), *Shifting Circles of Support: Contextualising Kinship and Gender in South Asia and Sub-Saharan Africa*, Sage, New Delhi, pp. 133–54.

Schneider, D. (1972) (originally published by University of California Press, Berkeley, 1951) 'Some Distinctive Features of Matrilineal Descent Groups,' in D.M. Schneider and K. Gough (eds.), *Matrilineal Kinship*, A.H. Wheeler and Co., Allahabad, pp. 8–27.

Sinha, A.P. (1971) 'The Pnar Family,' in S.K. Chattopadhyay (ed.), *Tribal Institutions of Meghalaya*, Spectrum Publications, Gauhati.

Syiemlieh, P.B. (n.d.) *The Khasis and their Matrilineal Systems*, published by the author, Shillong.

Synrem, K. (1992) *Revivalism in Khasi Society*, Sterling Publishers, Delhi.

Warjri, J. (1993) 'The Church and Women,' in *Report of the Meghalaya Church Leaders Conference*, Shandora Press, Shillong, pp.137–60.

Whitehead, A. (1981) '"I'm Hungry Mum": The Politics of Domestic Budgeting,' in K. Young, C. Wolkowitz and C. McCullagh (eds.), *Of Marriage and the Market: Women's Subordination in International Perspective*, CSE Books, London, pp. 88–111.

Yanagisako, S.J. and **J.F. Collier** (eds.) (1987), *Gender and Kinship: Essays Toward a Unified Analysis*, Stanford University Press, California.

X

Women and Forest:
A Study of the Warlis of Western India

INDRA MUNSHI

With the loss of access to forests, in the last century-and-a-half, the Warlis and other *adivasi* (indigenous peoples) of Thane district in Maharashtra state have lost not only an important source of livelihood, but also the basis of their religion and culture. However, many of their myths, practices and rituals have survived and provide interesting insights into the past. But the loss of status, both economic and social, is greater for women. From having been more or less equal partners with men in hoe cultivation and food gathering prior to the advent of the British in 1818, women have come to play a less important role in settled plough cultivation. Even more significant, property rights in land are restricted to men only. Evidence from indigenous people elsewhere in the country reveals the same story.

Depletion of forests has also meant a reduction in women's income from the sale of non-timber forest products (NTFPs). Since providing meals for the family is almost entirely women's responsibility, it has also resulted in greater hardships for women who must walk longer distances to fetch firewood, fruits, nuts, vegetables, and other necessities. Cattle grazing and collecting cattle manure are also very time-consuming for women and children. Collection of material for *rab*,[1] a traditional agricultural practice in the region, has become more arduous for women who are responsible for it.

In another respect, too, the loss of forests has had a negative effect on Warli women. The *bhagat* (the medicine-man-cum-priest) finds himself less effective due to the loss of medicinal plants. The old conflict between the *bhagat* and the *bhutali* (witch) recurs in a new context where the *bhagat* must find a scapegoat. The conflict between the witch and witch doctor, which can be traced to the early subordination of women, acquires a new reason for its existence, along with the contest for scarce land resources.

Women and Warli Culture

The transformation of the Warlis from shifting cultivation, hunting, and gathering to settled plough cultivation with some hunting and gathering has deepened the existing gender inequalities and weakened the position of women in the community. When and how the subordination of women occurred within the Warli community is not clear. However, despite the relative freedom of the Warli woman to move freely, select her partner, divorce, remarry, sing, dance, and drink with men as compared to her counterpart in caste society, inequalities exist in two crucial areas: political and ritual.

Women are barred from participating in the entire cycle of magico-religious ceremonies. They are not permitted to enter the precincts of the *raval* (training school for *bhagat*) for fear of causing the spiritual destruction of the *bhagat*. A woman cannot conduct the propitiation ceremony of the household gods, though she may attend the ceremony. On Diwali, when the household gods are taken out and worshipped, women are not permitted to touch or handle the idols. Women are considered impure and menstruation is seen as an outlet for their sins. Particular care is taken that no menstruating woman even casts her shadow on the images of gods. Women do not participate in the worship of the *vagh dev* (tiger spirit) who protects the village from external dangers. They cannot cook or partake of the sacrificial meat, nor can they cook or serve guests. This is only performed by men (Munshi, 1986). But the very fact that she needs to be controlled, subjugated as *bhutali*, suggests that there was probably a time in the history of the Warlis when the woman was more powerful. The part played by priestesses in marriage ceremonies, where the *bhagat* has almost no role, indicates that "there is likely to have been a period when the priestess as a representative of the mother goddess enjoyed a similar if not higher importance than the *bhagat*" (Dalmia, 1988: 46). On another important occasion, too, namely *zoli* (tying of the cradle), which initiates the newborn child as a member of the tribe and gives it a name, women play an important role. In the song of the corn goddess, Kansari, there is a suggestion of the context in which the subordination of the mother goddess, matriarchal culture, and woman might have occurred.

There is little information on the gender division of labor among the Warlis in the precolonial era. However, given the low level of technological development and the communal nature of production, it does not seem to have been clearly defined. Women participated in tasks connected with hoe cultivation, like preparing the field, weeding, and threshing, and were the primary collectors of fruits, vegetables, nuts, and small timber, while hunting was largely men's activity. Both women and men were engaged in minor trade of forest products. Mehta (1999) observes that among another *adivasi* group, the Koknas, in Thane district:

> women did more of the domestic and reproductive work and took on more of 'unskilled' manual and repetitive tasks in agriculture and food collection, while men did some domestic and agricultural tasks, and also hunting. Petty trade involved more men even though women also engaged in small trade of forest produce (Mehta, 1999: 70).

This was the situation within the overall context of male domination in the social and political domains.

Our fieldwork among the Warlis and other indigenous people in the region provides interesting information on hunting. It was reported that:

> Men go for *shikar*. Women don't go for *shikar*. They only kill animals if they happen to see them while collecting wood or other produce in the forest. Men hunt with bows and arrows, catapult, and sometimes with guns. Women do not use either. They use stones, sticks and sickle to kill. There is a taboo on women using the catapult. Men hunt rabbit, pig, deer, civet, wild

cat, iguana, squirrel. Women also kill small animals. Men go with weapons with the objective of hunting game whereas women go for other things and kill if they chance upon an animal.

The men have by and large forgotten the use of bow and arrow. On two special days, *sankrant* and *shimga*, all the men go hunting in the forest. There are no such hunting expeditions for women.[2] Since no large wild animals are present in the forest, both women and men kill small- and medium-sized animals. However, different terms are used for the same activity: "kill" when undertaken by women, and *shikar* (hunt) when undertaken by men.

A study of Birhor hunter-gatherers in Jharkhand shows that although there is a division of labor, it is not as well defined as in the agricultural communities. "While child care, food preparation and gathering are mainly the responsibility of the women; in hunting, rope-making and house-construction there does not seem to be much division of labor" (Kelkar and Nathan, 1991: 38). Birhor women help men in rope-making, but marketing the product in the neighboring agriculture communities is the women's responsibility. Women are the key persons for establishing contact with outsiders and conducting business (Sarkar, 1994: 54). The lack of a rigidly defined division of labor, Kelkar and Nathan observe, is a general characteristic of foragers everywhere (1991: 38).

The colonial land policy firmly established men's ownership and inheritance of land through legislation. Among the agricultural communities, as distinct from shifting cultivators and hunter-gatherers, the most important means of production is land, which is under men's control. Women are thereby excluded from ownership of private property. In most indigenous communities, a young girl has a right to maintenance till she marries, just as a widow has a right to manage and control her husband's land if she does not remarry or leave the village. Among the Warlis, women's claim to land is never supported. In earlier times, a *ghar jamai* (a man who moves to the wife's parents' home after marriage) was given a share in the land belonging to the father-in-law, but with growing pressure on land with each passing generation, the practice no longer exists. A childless widow is often accused of being a witch of having a love affair by her relatives, and chased out of the village in order to secure control over the land. The land issue is important for understanding the phenomenon of witch hunting in many *adivasi* communities at present.

In agricultural communities, a woman is not only excluded from ownership of land, but is also prohibited from the most crucial activity of production—ploughing. Among the Ho and Munda, a woman must not even touch the plough. Among the Santhal, Oraon, and Kharia, there is no taboo on women touching the plough or roofing the house. Kelkar and Nathan (1991) point out that key taboos such as ploughing and roofing the house, although justified in terms of religious beliefs, do play an important role in establishing and maintaining women's subordinate position. They ensure that women remain dependent on men for crucial activities.

Although going to the forest with friends is still an enjoyable activity for most Warli women, having to walk long distances with heavy loads and being harassed by forest guards take away most of the pleasure. When asked if they had fun in the forest, most women replied, "Where is the time to have fun?" One woman said, "Sometimes we take a bottle of liquor in our basket, and drink together in the forest." The forest as a social space where women went in groups to collect firewood, gossip, eat, sing, roam, bathe, and escape from the drudgery of household chores, has shrunk over time. It is no longer the familiar space they visited regularly, whose plants and animals they knew. Although *adivasi* women still go to the forest, it is nevertheless a degraded and hostile space, not the place that symbolized freedom and abundance. Rituals associated with the forest will soon disappear because the gods of the forest, *hirva* (the forest spirit), and *waghobha* (the tiger spirit who protects the villages of the Warlis, Koknas, Thakurs, Katkaris, Kolis, Bhils), and a large number of sacred trees, animals, and birds are also disappearing.

Gathering of food and medicinal herbs requires a fairly elaborate knowledge of plants, trees, and roots. During our interviews, we found that many women did have the necessary knowledge and, in fact, were more knowledgeable than the men. Women treat a number of common ailments at home. The *suin* (midwife) knows a lot about medicinal plants useful for childbirth, abortion, and related problems. Depletion and destruction of forests and restriction of access to them meant the destruction of the material basis of such knowledge. A number of Warli women still have this knowledge, but it is fast being eroded by forces of modernization like the entry of modern medicine and commercialization of forest products. In the context of reduced supply of herbal medicine and restricted access to it, as well as the pressure on land and its control by men, witch hunting acquires a new significance.

Myth and Reality of *Bhutali*

Bhutali are believed to possess special powers by which they can instantly put a person to death or cause illness. They utter evil charms which immediately affect the victim, no matter what the distance between the witch and the victim may be. This special power is known as *mooth mavane*. The *bhutali* can approach her victim in the form of a fly, a cat, or a hen. She can move around invisible, and cover long distances. She generally moves naked at night. In 1940, it was observed that Warlis had tremendous faith in the power of the witches whom they feared very much. Interestingly, it was reported, the Warli *bhagat* believed that Sukra (Jupiter) was once burnt and some of the luster of that brilliant planet carried away by the *bhutali*, and that is what made her so powerful (Save, 1945: 66–67). A Warli *bhagat*, the priest-cum-medicine-man, had the following explanation for the origin of *bhutali*:

God sent many diseases to mankind by which creatures in this world die. He was afraid people would not call him kind and good, if they come to know that it was he who sent miseries to the world. They would curse him and call him bad names. In order to avoid all such blame, God created *bhutali* so that people attributed their sickness and miseries to her.

Later God also desired that a *bhutali* should be easily marked out, so he wanted her to have horns on her head. But the idea had to be dropped since it turned out that God's own mother was a *bhutali* (ibid.: 1945: 67). The story clearly suggests that all of womankind is untrustworthy and suspect since any woman can turn out to be a witch.

Another story that traces the origin of *bhutali* is often recounted by women, probably in response to the earlier one. The god Naran Dev, wanting to prove his superiority over Kansari (the corn goddess), once suggested that they transplant their fields separately rather than jointly as they always did. He cut the rice seedlings into halves, keeping the lush green shoots for himself and leaving the dirty roots to Kansari. The transplanting done, Naran Dev's fields looked lush. He was confident of humiliating Kansari with his beautiful harvest. But after some time, his shoots wilted and Kansari's roots brought forth a bountiful crop. Naran Dev was unable to understand the reason for this catastrophe. He accused Kansari of being a witch and destroying his crop.[3]

The dark goddess, Kansari, is clearly a symbol for the coarse grain *nagli* which was commonly cultivated by the Warlis on the hills slopes by shifting cultivation before the transition to plough cultivation. Warli songs suggest the subordination of the goddess Kansari by god Naran Dev. She is humiliated by the gods and thrown out of their kingdom, although at the same time the farmers and the gods realize that they cannot do without her (Save, 1945). In another version, the gods, the chief of whom is Indra, are engaged in a constant struggle with her to capture and attain her (Dalmia, 1988). The supersedure of the mother goddess probably coincided with the establishment of patriarchal norms in a situation where they did not exist. One version of the Kansari story reads as follows:

God made the earth, the mountains, wind, the sun, the moon and all creatures big and small. There was, however, no food for these creatures. Naran Dev prepared a plot for sowing 18 seeds. He meditated upon the mother corn and went deep into the darkness in search of her. He arrived at the city of *dhanol* (full of corn), but mother corn refused to accompany him to the assembly of gods.... [Eventually,] he persuaded her to go with him.... The gods requested her to give them food as a means of livelihood. At last she, the mistress of 18 corns, consented to create corn. Thereafter gods got abundant crops. But they soon got tired of the corn and said, 'We do not wish to have this dark beauty (*nagli*) in our kingdom. Find a husband for her and drive her out! (Save, 1945: 174).

The insulted corn deity immediately leaves the gods' abode. The gods, however, soon find that without her they are reduced to starvation. Subsequently, Naran Dev chases Kansari and brings back some drops of her blood. He returns it to the farmer who, he feels, would take better care of the corn. In still another version, there is a persistent lament about the loss of Kansari who is constantly angry with both the farmer and the gods, but more with the gods, particularly Indra. The Warli songs lament their abandonment by the mother goddess: "Kansari the great mother has abandoned us. Water was let loose, the field dried up" (Dalmia, 1988: 86). But the superiority of the god Naran Dev over the goddess *Kansari is well established in* songs sung in the new rice ceremony (*nava bhat*). Naran Dev is the supreme god.

An assembly is held at the palace of mother Kanu.
But Naran Dev is managing everything.
In the front yard of Kansari is held an assembly.
But Naran Dev is doing everything
(Save, 1945: 153).

Recent fieldwork in Thane district revealed that *bhutali* hunting
continues to occur frequently in the region. As one political activist
observed, when there is an illness, or death among human beings or
cattle, widespread crop disease or failure, the *bhutali* is held responsible. In such a situation, the villagers go to the *bhagat* to identify
the cause of the calamity. Very often, the women of the village are
collected and warned that unless the harm is undone, action would
be taken against the witch. If the situation does not improve, the
bhagat goes ahead with the task of identifying the *bhutali*. Through
a variety of rituals like *dan herne* (reading the message in grains of
rice), *diva herne* (identifying the witch in the light of a lamp), and *vati
chalavne* (using a cup which moves and identifies the witch), the
bhagat tries to find the witch. The *bhagat* who identifies the witch is
usually not from the same village. Through careful questioning, he can
locate women who are socially weak and vulnerable, quarrelsome,
destitute, with poor family support, midwives, and so on. He gives a
vague description of the woman or women who caused harm.

The next step is to find the woman. The men of the village call for
an identification parade. Women are sometimes made to stand on a
hot pan as the Warli believe a witch's feet do not burn. At other times,
the witch is simply pointed out by someone. Another way is that all
the women who fit the description are beaten till someone confesses
to the crime (Kashtakari Sanghatana, 1984). Once the witch or
witches are identified, the whole village goes through the ritual of
exorcising her. The woman is beaten, sometimes to death. No one,
not even close relatives, protects her from the attacks, for fear of
being accused as an accomplice. A survivor in most cases leaves the
village. If the woman dies, she is buried rather than burnt, and the
villagers give no information regarding the death to the police.

Bhagat and *Bhutali:* Contest for Knowledge

The chief function of the *bhagat* is to detect the witch and to devise
some means of getting rid of her. Both try to outwit each other. The

bhagat is an enemy and the witch tries to undo the work of the *bhagat*. In fact, during his training, the *bhagat* shuns any contact with women lest they should turn out to be witches and usurp his power. Within the belief system, men are said to possess the power "granted to them by the gods" to identify witches.

As the witch-finder, the *bhagat* is the greatest enemy of the witch. For the same reason he enjoys a very high status in the community. From all evidence available, it appears that the struggle for power and superiority between the sexes is manifested in the antagonism between the *bhagat* and the *bhutali*. Knowledge is contentious, and each tries to undo the other by acquiring the knowledge of the other. The *bhagat* "knows all the evil charms equally well as a witch. A powerful *bhagat* can undo what a witch has done. He can thus render the *mooth* of *bhutali* null and void" (Save, 1945: 67). Her discomfiture and fear of the fact that he is thoroughly conversant with her *vidya* (knowledge) is brought out in the following story. On realizing that the *bhagat* was more than a match for her, the *bhutali* prayed to God and made a vow to offer him a goat if he asked the *bhagat* not to reveal her secrets, particularly her name, to anybody. God granted her the wish. She was satisfied that the *bhagat* did not reveal her name to anybody. But the *bhutali* did not keep her promise and, instead of offering a goat to God, she offered a louse from her head. God was angry and cursed her that in her next life she would hang upside down from a tree and wander only at night.

That the *bhagat* and the *bhutali* constantly try to acquire each other's knowledge, and thereby power and superiority over the other, is obvious from the secrecy maintained during the training for *bhagat*hood. Every year an established *bhagat* initiates a few pupils. The training lasts about four months, conducted generally at night. During this period, the pupils are forbidden to eat certain foods, but, most importantly, are not allowed to be close to a woman or to have sex. The training or *raval* is held in a secluded spot outside the village in a small booth constructed for the purpose. No women are allowed there. In fact, the booth is near a water source for the specific purpose of drowning out sounds which women may otherwise hear. Even the sound of a woman's bangles must not be heard by those in the *raval* (Dalmia, 1988).

As one Warli man put it, "The education for *bhagat*hood is forbidden to women because they become witches and misuse the knowledge. In spite of this the women sit out of sight and without the knowledge of the *bhagat* memorize everything" (Dalmia, 1988: 46). A large area around the school is enclosed by a boundary line on which wooden and iron nails are fixed and sand is poured. This boundary is meant to serve as a barrier to witches. Inside the boundary lies the area which is under the charm of the *bhagat*. If a witch tries to cross the boundary, she would have to take out the nails fixed in the ground and count every particle of sand, which would naturally take a long time. The witches, however, are only too eager to attend the *raval* unseen to watch the proceedings and create hindrances in the way of the *bhagat*. An ordinary person may not feel the presence of a witch, but a powerful *bhagat* feels it and chases her away.

In her book on Warli painting, Dalmia describes a painting which depicts the paranoia the *bhagat* feels about witches. In the painting the *raval* is in the center, with the chief *bhagat* sitting under the shed. The pupils are seen with whips to beat themselves with. Some *bhagat* are running after witches with whips. The witches are shown with flying hair rather "like the mane of a lion." "Unwittingly," Dalmia observes, "they seem to appear stronger than the *bhagats* chasing them." One witch is shown riding a dog to reach the *bhagat*. Another witch sits on a bundle of grass, trying to eavesdrop, and may have ridden the grass for she "comes like the wind" (Dalmia, 1988: 46).

The following Santhal myth highlights how, through trickery, women acquired the knowledge they had no right to. As the story goes, the village men once approached Marang Buru and complained about the insubordination of their women, and begged him to teach them how to keep their womenfolk in order. When the women came to know about this, they got their menfolk drunk, wore their clothes and tricked Marang Buru into teaching them. He taught them incantations, which gave them the power of eating men. The next day, when the men came, Marang Buru realized that he had been tricked. He then made the men "experts in the art of witch finding" (Kelkar and Nathan, 1991: 97).

On the other hand, the *bhagat's* superior power is due to the fact that he is not only familiar with *bhutali vidya* but also has the knowledge to fight witchcraft, by the use of material and magical medicines. He is often a medicine man, herbalist, and diviner all in one. Speaking about the African witch doctor, Parrinder (1958) observes that he is usually both a controller of witches and a dealer in magic and herbalism. As a diviner, he is able to detect the witches, and as a magician, he counteracts their evil powers with his medicines and magical spells.

In his article on witch-finding in Zambia, Yamba confirms this. There is an unclear distinction, he observes, between witch-finder and traditional healer in the minds of the local people. A traditional healer's expertize may include activities such as "driving out ghosts and bad spirits," in other words, witch-hunting (Yamba, 1997: 205). The *bhagat* also combines knowledge of medicine and magic in combating the *bhutali*, but we will discuss this in greater detail later.

The *bhagat* tries to get access to the *bhutali vidya* in many interesting ways in order to be most effective as a witch-finder. We have the story of a Warli *bhagat* who married several women, many of them *bhutalis*, just to know the charms used by them so that he could work effectively against them. The old *bhagat* said he preferred a *bhutali* for his wife simply because she could tell him what particular evil charm she practiced on a person and when the influence of that charm would end. He could thus equip himself with very valuable information from his wives. They practiced *bhut* (witchcraft) and he provided a cure for it (Save, 1945).

The association between magic and medicine is well established. Both the witch and the witch doctor combine knowledge of magic and herbal medicine. In a very different context, it is observed that in Europe, while many lonely old women were suspected of witchcraft, some of them were wise women who were experts at making herbal remedies (Parrinder, 1958). These remedies were sought after during illness, but might be held against them as witchcraft if the patient died. Recent writings reveal that in 15th, 16th, and 17th century Europe, women were persecuted because they had special knowledge of healing and midwifery that challenged the new all-male professional doctor.

The competition and conflict over healing took a particularly savage form in Europe in the 16th century. Ehrenreich and English (1978) tell us that the conflict between women's traditional wisdom and men's expertize centered on the right to heal. Healing had traditionally been the prerogative of women in general except for the very rich. The women who distinguished themselves were not only midwives, but herbalists and counselors serving women and men alike. Witch hunts were almost exclusively for peasant women, and among them lay women healers were singled out for persecution. The inquisitors, the authors point out, directed their greatest wrath towards the midwife, asserting, "The greatest injuries to the Faith as regards the heresy of witches are done by midwives; and this is made clearer than daylight itself by the confessions of some who were afterwards burned" (ibid.: 36). Throughout the witch hunts the church lent its authority to the doctor's professionalism and denounced nonprofessional healing as heresy. In North America, however, the woman healer was not eliminated by violence. She was defeated in a struggle, by repression and slander, when healing was transformed from a "neighborly service" to "a commodity and source of wealth in itself," a man's enterprise (ibid.: 41).

Interviews with the Warli women suggest that many of them had a fair knowledge of the medicinal properties of plants, trees, roots, and herbs. In fact, they treat many common ailments at home. Warli women did and still do have this knowledge, which is being fast eroded by forces of modernization like the entry of the doctor and commercialization of forest produce. In fact, as one woman observed, "Women give medicines only for small ailments because they are afraid that if anything happens they will be accused of being *bhutali*." One is tempted to ask if there was a time in the history of the Warlis when the *bhagat* saw women experts like the *suin* as a threat to his position of domination and found a way of subjugating them by denouncing them as *bhutali*, the evil ones.

The *bhagat* has long enjoyed and continues to enjoy, very high status as the medicine man. Till recently, the *adivasi* depended totally on the *bhagat* for the treatment of illness. Even now, given that the government medical service is inadequate and private service too expensive, most people continue to go to the *bhagat*. That the *bhagats* had access to and knowledge of powerful herbs is illustrated by the story of a *bhagat* who lived a century ago. He is believed to have had such "powerful herbs with him that even a bullet could not hurt him. His corpse could not be burnt unless the herbs were taken out of his body" (Save, 1945: 72). That the efficacy of the *bhagat* is progressively diminishing for a number of reasons, including deforestation and the loss of traditional herbs, could have some bearing on the increased instances of *bhutali* at present.

The point being made here is that, even though as one activist from the Sanghatana suggests, the functions of the *bhagat* and the *suin* are distinct in the spheres of curative and reproductive health, there was probably some competition between the two. The mystery and the fear associated with menstruation and childbirth could well have been translated into mysterious and fearsome powers of the person dealing with it. The transformation of the *suin* into *bhutali* is not unimaginable.

Land, Forest, and the Phenomenon of *Bhutali*

Recent research shows that witch-hunting in many *adivasi* communities in India is linked to contested land rights. In the Jharkhand region, it was found that witch-hunting was related to the attempt of male agnates to remove the threat to their property rights posed by the widow's life interest in the land (Kelkar and Nathan, 1991). Mehta (1999) cites instances from Thane district where men accuse women of immoral behavior, or even witchcraft, to pressurize them into renouncing their rights to land. Most of the time, the accused women were old, widows, or women with little family support (Jena, 1994). Our fieldwork supports this observation.

A recent interview with one such woman provided important insights into the phenomenon. The Warli woman, approximately 45 years old, had the following story to tell:

> My husband drowned a few years ago when the fishing boat on which he worked sank. I had a two-year-old son at that time. My husband's nephews wanted to take away the land which my husband owned. They called me a *bhutali* and accused me of having killed my husband. They tried to molest me and drove me away from the house and the village. I came with my son to my mother's house . I will go back when my son grows up and claim the land.

Interestingly, she suspects that her envious sister-in-law, a *bhutali*, cast an evil spell on her husband resulting in his death. She does not allow her son to come into close contact with or eat and drink with her husband's family.

In Chota Nagpur and Santhal Parganas, it is reported that superstitious beliefs are merely used as a cover by male relatives to capture the lands of widows, deserted women, or single women. In Chaibasa court, many such cases are pending where single women, unmarried or widowed, were accused of being witches and killed by their own male relatives. But the real reason was to eliminate the women and take away their land (Manimala, 1988). There is an increasing tendency to kill whole families. In Singbhum district, a woman who was accused of being a witch and punished reported, "All this was done just to snatch about four *cottahs* of land that was in my possession" (Zahir, 1997: 62). Pahadia women face a different kind of harassment. They are thrown out of their villages by their male agnates, who conspire with the village chief only to get hold of the land. Such women often live together outside the village in small groups, but continue to be harassed by the village chiefs, policemen, and forest contractors (ibid.: 23). Among the Santhals the accused woman is fined an amount ranging from Rs 500 to Rs 5,000, which is shared between the village chief and the witch doctor. If she refuses to pay, she is beaten, or even stoned to death (Bannerjie, 1984).

Activists of Kashtakari Sanghatana (KS), a political organization working in the region since 1976, report increased incidents of witch-hunting in recent years. In a private communication one activist observed that these days *bhagats*, few in number, do less herbal treatment and more *bhuta* (dispelling of the evil spirit). Therefore, more *bhutalis* have to be found. The government medical service is inadequate and private service very expensive. Most people find private doctors and hospitals too expensive and prefer to go to the *bhagat* for treatment. "He is accessible, he is known, he understands, he is inexpensive, he is reliable, and he is acceptable ... he forms an integral part of the *adivasi* healing system" (Kashtakari Sanghatana, 1984: 90). But the efficacy of the *bhagat* has been reduced with defor-estation. Most of the traditionally used herbs are difficult to find, and many *bhagats* have decreased the use of herbs, because finding them is time-consuming. Since handing over knowledge of herbs means the *bhagat* loses his power and efficacy, many *bhagats* do not hand down the knowledge acquired over the years.

In addition, with migration to slums on the fringes of the cities, many *adivasi* return with new diseases the *bhagat* is unable to cope with. Added to this is the fact that lack of nutrition from the forest has led to a lowering of the general health of the *adivasi*. Therefore, increasingly, the cause of illness (and death) has to be fixed on the witch and the cure found in witch-hunting (Kashtakari Sanghatana, 1984: 90). An old Warli man told me, "We don't have the kind of knowledgeable *bhagats* we used to have. They knew about herbal medicines, used to go to the forest and collect herbs. Now you don't get such *bhagats*."

An idea that underlies most feminist analysis of witchcraft is that the accused women were more often than not single women, outside the control of the patriarchal family. The threat and terror of an accu-sation of witchcraft were used to contain them. The fact that the *bha-gat* finds himself less effective means that he must find a scapegoat in the *bhutali* for illness and death. He, like the *badawa* of the Bhils, still enjoys a high status in the community as a witch-finder (Varma, 1978). Through a ritual, the cause of illness/ death is detected and the witch identified. The pent up aggression and anger of a deprived existence find violent expression against an often helpless woman. Having removed the source of "evil" and harm, order and normalcy are restored, at least for a while.

In the case of the Warli women, this may also be linked to the high status they enjoyed in the past. Witch hunting, an attack on that status, changed the social relations in favor of men. Its continuation becomes necessary to maintain the patriarchal order, by constantly putting down all insubordination and challenge from women. Men's fear of women's power has not been overcome and they still see a threat to the patriarchal order. The object so feared evokes the most violent reaction, leading them to seek in extreme situations even its annihilation. Witchcraft accusations are therefore "integrally linked to and served to reinforce and reconstruct—the male status quo" (Hester, 1998: 301). However, the Warli women are taking on the combined might of institutional patriarchy and the men within their community in their fight to preserve their forests.

Warli Women's Voices

The conversion of natural forests into commercially valuable monocultures, the overall depletion and degradation of forests, and the restrictions on *adivasi* use and control over forests (Munshi, 1995: 85–111) have resulted in the loss of a crucial source of subsistence for all *adivasi* in the district. For both women and men, it has meant greater hardships in the collection of small timber for house construction or agricultural implements. Women face additional problems in collecting fuelwood, material for *rab*, fruits, roots, and vegetables essential to tide over the lean period or supplement a meager diet, herbal medicine for ailments, thorny bramble for hedges, grazing cattle, and collection of cattle droppings for fertilizer. Women's burden has increased greatly. To quote the Warli women themselves:

> We don't get wood for cooking. One has to cut wood out of sight of the forester. One is always afraid. If any khaki uniformed person comes, we run for our lives. If anyone cuts a tree for a plough, and a forester comes, he has to leave his axe and run. The forester takes it away. There are so many restrictions that it has become difficult to get wood for fuel. When *adivasi* women go to the forest, they are threatened and chased by the forester.

Earlier, a lot of roots were available. *Adivasi* used to eat roots and satisfy their hunger. Medicines were also available. If a woman fell ill, other women would go to the forest and bring medicine from the forest. Now that is not available. Earlier there were many trees that provided medicines. Timber Board has finished all of them. Earlier there were no doctors or hospitals. For a number of ailments people got medicines from the forest, they didn't go to the doctor. There were women who were like doctors, they were knowledgeable about medicines. Now even if you roam the forests for five days, no medicines are available. The forests are destroyed, no medicines, no roots are available.

For people of the forest, wood is a big gift of god. But if we don't get fuelwood, how will people cook and eat? This question is in everyone's mind. Timber Board has grown Nilgiri and acacia everywhere, for which they have destroyed other trees. Nilgiri or acacia are not useful as fuelwood. Those who have money can cook on gas stoves. What will those who do not have anything do?... If the forest department does not allow women to bring wood from the forest, will they eat raw rice? What will they do?... We cannot eat uncooked rice.

It is important to know that in the case of the Warli, the forest not only provides timber and a large variety of non-timber produce essential for survival, but also material essential for agriculture. *Rab* continues to be a bone of contention between the forest department and the peasants. But peasants, primarily *adivasi*, continue to practice it not only because it is time-tested and cost effective (since women and children collect material for *rab*), but because they cannot afford alternative methods of cultivation with irrigation, hybrid seeds, chemical fertilizers, and pesticides.

Warli women substantiate the argument that deforestation and degradation of village commons have adversely affected rural poor households and women members of such households in particular (Agarwal, 1991; Fernandes and Menon, 1987). Agarwal points out that firewood, the single most important source of domestic fuel in rural South Asia, is mostly gathered and not purchased. It is estimated that in six states of India, in the 1990s, primarily women and children spent an average of three–four hours and traveled 4–5 km for firewood each day (Agarwal, 1997: 12). Not only has the collection, sale, and barter of firewood become more time-consuming, arduous, and risky, it has also meant a decline in an independent source of income for the indigenous women.

In both nonindigenous and indigenous poor peasant households, it is the women who gather fuelwood and NTFPs from forests and village commons. Indigenous women in particular are major gatherers of NTFPs for consumption and sale. An estimated 70 percent of NTFPs are collected in the indigenous belts of five states in India. Men, in contrast, tend to draw on the commons much more for timber, including small timber for agricultural implements and house construction (Agarwal, 1997: 11). Interviews with Warli women in several parts of the district revealed a growing sense of alarm over the rapid depletion of the forests. Many of them reported having to walk three to four hours and even more to collect firewood. Most women spend the morning in the forest. While roots, berries, fruits, leafy vegetables, and firewood are collected for consumption, seeds, resin, leaves, and firewood are bartered for onions, garlic, fish and spices.

Women from landless households go to the market every day to exchange NTFPs for daily necessities. Occasional cash income is spent on trinkets, bangles, and clothes for the children. For households with some land, this not only supplements the daily diet, but also constitutes a major source of food and income during the lean season which lasts from four to six months. During the "difficult months," March to May, most *adivasi* subsist on *paje* (a watery gruel made from rice) and *kand* (a bitter edible root from the forest). The root must however be cooked for a long time before it can be eaten.

Jungle Bachao: Women's Role

In recent years, *adivasi* women and men have shown a good deal of initiative and interest in forest protection, supported by some political organizations. In some villages, women lead the *Jungle Bachao* (Save the Forest) movement. Groups of women patrol the forest. Although it might be too early to say, *adivasi* women and men seem to show different approaches to forest protection. Whereas men want to protect timber, women want to protect the forest. Women have destroyed teak saplings planted by the forest department in the region. Illegal trade of timber is thriving in the district and *adivasi* men are employed to fell the timber and transport it to the road, from where it can be transported by trucks. They are paid a small sum of money for their labor. This trend is noticeable in other parts of the country, where local communities are becoming conduits to serve urban and commercial demands. In other instances, however, women have opposed the theft of timber. They know that forest depletion makes life harder for them. In the mid-1990s, angry women chopped timber logs stacked by the roadside in Sukhdamba village and carried them away to use as firewood.

To protect their forests, *adivasi* women activists in some villages took the lead, attended meetings, and participated in decision-making. They often met women from neighboring villages mobilizing them for the *Jungle Bachao* movement. Groups of five men would generally patrol at night, while groups of 15 women kept watch during the day. Subsequently, more women than men kept watch over the forest. If anyone was caught cutting timber, they tried to convince the person to stop, and; in some cases, even punished her or him.

Interviews with a number of *adivasi* women revealed their interest and involvement in protecting the forest. "Women need forests more for fruits, root, fuelwood, leaves for *rab*. They need herbs for medicines. That is why they are so active," they said.

> *Jungle Bachao* is very good. Because all the people support and help, the forest has been saved. Women leave household work and go for forest protection. Men patrol at night. Thieves are scared. We get wood and leaves because the forest has been saved.

> Our people are saving the forest so it has been protected. Women also patrol the forest. Forests are needed. If they are there, our grandchildren will get timber. Otherwise even for cremation there will be no wood. There will be no leaves for *rab*. Women, men and children all help in protecting the forest. It is good. Forest department has cleared the forest flat. If we keep it, it will grow.

> Women believe in saving the forest. The forest nearby has been saved, it is good for the *adivasi*. If the forest is saved, only then the *adivasi* can be happy.... If there is no forest, no one will live there.... As the forest decreases, the sorrow and difficulty of the *adivasi* will increase.

In villages where the *Jungle Bachao* movement has not started, there is concern.

Our forest is not protected. The government sees big trees and takes them away. If the forest is not there, it is bad for the *adivasi*. Even now there is less wood, if it is further reduced how far will we go? Forest should be saved and grown but who will do this? Who will tell the government?

It must be noted that in the responses and action of the *adivasi* women, I do not read an inherent ecological consciousness special to women, nor do I wish to essentialize *adivasi* women's closeness and concern for forest/nature. The *adivasi* women and men share a long history of close association and dependence on the forest, and this is reflected in their material and symbolic reality. Their responses only highlight the fact that the destruction of forests has adversely affected the community, especially women, given the nature of their work and activity in the forest. As Beate Martin puts it,

Gender determined work segregations, at home and outside, lead women to interact with the local resource base far more extensively than men, which in turn results in the differential impact of resource degradation on women and men (Venkateswaran, 1995: Foreword).

Women are victims of larger processes beyond their control, but also agents for change which are in consonance with their interests, and which represent the long-term interests of the community. The *adivasi* women's initiative may be seen as an attempt to redefine and reestablish norms for the management and use of resources by the community, given that the old norms have broken down in the face of rampant commercialization and overexploitation in the last nearly a century-and-a-half. In this, as they themselves suggest, they have the support of their men.

Conclusion

Women's struggle for access and rights to resources would have to include the right to land. Given that there is little support from the community on this issue, few political organizations have succeeded or even attempted to take up this issue. Hence, the Warli women's struggle for a dignified existence, including their rights to material and social resources, is a struggle against the patriarchal norms and values of both the state and their own community which is manifested at various levels.

The old practice of *bhutali* continues in the new context of increasing alienation of the *adivasi*, the Warlis, from their natural environment, their marginalization, and incorporation into the mainstream society at its lowest level. It is imaginable that, in a situation of extreme stress resulting from the transformation of the community from a relatively independent subsistence level existence to one of deprivation and exploitation, the aggression is directed towards the most vulnerable section of the community, the women.

Given, however, the widespread nature and the deep-rootedness of the belief, the political organizations working in the region are not only cautious but also ineffective in making a positive intervention in putting an end to the practice. Most reports on recent witch-hunting record the inability of the non-government organizations (NGOs) working in the *adivasi* areas to control the phenomenon. For fear of alienating the community, most organizations have refrained from taking up the issue. In most cases, the bureaucracy and the police are not effective either.

Recently, however, 26 women accused of being witches, ostracized and persecuted by their own people, gathered in Patna under the banner of "*Dain* Atrocities and Measures to Prevent the Same," convened by an NGO called Free Legal Aid Committee. Alarmed by the incidents, the district administration launched a campaign against this practice. The NGO plans a statewide campaign against witch hunting (Zahir, 1997). The Kashtakari Sanghatana working in the region rightly sees the need to awaken the women against the practice, as part of women's larger struggle for emancipation and equality. The other necessary conditions are improved health facilities, less superstitious attitudes among the *adivasi*, and the involvement and education of the *bhagat* for more effective health care.

Given the many levels which the *bhutali* phenomenon permeates in the *adivasi* community, it is not easy to stop it. What is clear is that the Warli women's struggle for a dignified existence must include struggles against the state for rights to livelihood and resources, against the patriarchal norms and practices of their own community which deny them rights to land, and equality in the cultural/religious spheres.

NOTES

1. *Rab* is a system of seedbed cultivation, where a heap of twigs, leaves, grass, wood, and cow-dung is set on fire with earth on top to keep it down. When the rains come, seeds are sown and the seedbed ploughed lightly and harrowed. After an interval of 18 to 20 days the seedlings are transplanted to the field. The practice was seen as "wasteful" by the British, although agricultural experts recognized the usefulness of *rab* (Munshi, 1990: 433–43).
2. Many indigenous communities have prohibitions on women hunting or even handling weapons. Among the Santhal the bow and arrow is prohibited to women. In an interesting ritual, *jani shikar* (women's hunt) among the Munda and the Oraon people in Bihar, women dress in men's clothes, carry bow, arrows and axes and take part in a mock hunt. The gifts collected during the hunt such as fowl, goat, sheep, and pig are considered women's property.
3. I am grateful to Shiraz Balsara of the Kashtakari Sanghatana, a political organization working in Dahanu since 1976, for this story.

REFERENCES

Agarwal, Bina (1991) *Engendering the Environmental Debate: Lessons from the Indian Subcontinent,* CASID Distinguished Speakers Series No. 8, Center for Advanced Study of International Development, Michigan State University, Michigan.
———— (1997) 'Environmental Action, Gender Equity and Women's Participation,' *Development and Change,* 28 (1): p. 144.
Bannerjie, Indranil (1984) 'The Branded Witches,' *India Today,* 31 October.
Dalmia, Yashodhara (1988) *The Painted World of the Warlis,* Lalit Kala Academy, New Delhi.
Ehrenreich, Barbara and **Deirdre English** (1978) *For Her Own Good,* Anchor Books, New York.
Fernandes, W. and **G. Menon** (1987) *Tribal Women and Forest Economy: Deforestation, Exploitation and Status Change,* Indian Social Institute, Delhi.
Hester, Marianne (1998) 'Patriarchal Reconstruction and Witch Hunting,' in Jonathan Barry, Marianne Hester and Gareth Roberts (eds.), *Witchcraft in Early Modern Europe,* Cambridge University Press, Cambridge.
Jena, Kalandi (1994) 'Progress of the Educated Tribal Women,' *Man in India,* 74 (1): pp. 81–86.
Kashtakari Sanghatana (1984) 'The "Bhutali" Phenomenon: Why Are Women Hunted Down as Witches,' *Socialist Health Review,* 1 (2): pp. 87–92.
Kelkar, Govind and **Dev Nathan** (1991) *Gender and Tribe: Women, Land and Forests in Jharkhand,* Kali for Women, New Delhi.

Manimala (1988) 'Witch Hunt,' *The Illustrated Weekly of India*, 1 May.

Mehta, Mona (1999) *Suppressed Subjects?* Institute of Social Studies, The Hague, The Netherlands.

Munshi, Indra (1986) 'Tribal Women in the Warli Revolt 1945–47: Class and Gender in the Left Perspective,' *Economic and Political Weekly*, 21 (17): WS 41–52.

———— (1990) 'The Political Ecology of Traditional Farming Practice, Thane District,' *Journal of Peasant Studies*, 17 (3): pp. 433–43.

———— (1995) 'Where Have the Forests Gone? An Exploration into Thana District,' in Manorama Savur and Indra Munshi (eds.), *Contradictions in Indian Society*, Rawat Publications, Jaipur, pp. 85–109.

Parrinder, Geoffrey (1958) *Witchcraft*, Penguin, Great Britain.

Sarkar, Sampa (1994) 'Status of Tribal Women in Three Socio Cultural Dimensions,' *Man in India*, 74 (1): pp. 49–57.

Save, K.J. (1945) *The Warlis*, Padma Publications, Bombay.

Varma, S.C. (1978) *The Bhil Kills*, Kunj Publishing House, New Delhi.

Venkateswaran, Sandhya (1995) *Environment, Development and the Gender Gap*, Sage Publications, New Delhi.

Yamba, C. Bawa (1997) 'Cosmologies in Turmoil: Witch Finding and Aids in Chiawa, Zambia,' *Africa*, 67 (2).

Zahir, Naved (1997) 'Witch Hunting,' *Sunday*, 28 December.

XI

Empowerment and Disempowerment of Forest Women in Uttarakhand, India

꧁ꕥ꧂

MADHU SARIN

What factors trigger changes in women's position in a patriarchal forest-based society? How effective are devolution policies rhetorically committed to empowering women to participate in forest management? This chapter explores these questions with the help of case studies spread over nine districts of Uttarakhand (now the major part of the newly constituted state of Uttaranchal in November 2000) in north India. Beginning from colonial rule in the 19th century, 67 percent of the region's uncultivated commons, critical for sustaining its subsistence based agro-pastoral livelihood systems, have progressively been appropriated by the state as "forests." The accompanying reduction in villagers' forest access in a context where the use of forest resources has traditionally been considered "women's work" has had a profound impact on reshaping traditional gender relations. More recently, a host of state policies have attempted to promote women's socioeconomic and political empowerment. These have included normative interventions such as the reservation of one-third seats for women in *panchayati raj* institutions for local self-government and promotion of all women *van panchayats* (elected councils for managing village forests), combined with more innovative government programs such as Mahila Samakhya (women's empowerment) focused on mobilizing women for self-empowerment.

The recent policy intervention for devolution of forest management through village forest joint management (VFJM) introduced through a World Bank-funded forestry project has been overlaid on the existing and ongoing state initiatives. The forestry project also has a stated commitment to specifically target women and the poor as project "beneficiaries." In addition, there are several other government and donor funded projects (such as watershed development, water resource development, diversification of agriculture, etc.), each with its own component and strategy for enhancing women's participation. All these interventions have been introduced in a regional context which has a long history of people's own struggles for rights over land, forests, and water of which the Chipko Movement of the 1970s is the most famous. More significantly, village women themselves have been organizing their own struggles within their households and communities to gain greater control over their forests, as it is their daily lives which are most intimately impacted by forest quality, their proximity to the village, and institutional structures determining forest access.

This chapter begins with a brief overview of the historical evolution of the complex diversity of tenurial regimes for common lands and community institutions for their management in Uttarakhand. The second section examines local struggles to gain greater control over forest use and management, and of the forces aiding and hindering such struggles within existing formal as well as informal community institutions. In the third section, case studies undertaken to examine the impact of devolution policies for forest management on creating space for local control over decision-making, forest quality and management for enhancing livelihoods are discussed. Due to women's continuing centrality in the use and management of forest resources in Uttarakhand, the case studies focus on the impact of devolution policies on gender relations and women's spaces for forest management. The final section examines the impact of the recent forest devolution policy of VFJM, and the promotion of women's representation in *van panchayats*.

The Context for Forest Management in Uttarakhand

Nestled in the Central Himalayas, Uttarakhand is an area with rich forests and unique biodiversity. According to the 1991 census, Uttarakhand's population was 5.93 million, 78 percent of which lived in rural areas. The rural population of the hill areas is even higher at over 90 percent. While the area had an economy of abundance at the time of colonial occupation in the early 19th century (Guha, 1989; Nanda, 1999), today an estimated 45 percent of its economically productive workforce is employed outside the region for lack of local employment. While only 12.6 percent of the hill region's area is officially under cultivation (Saxena, 1995), rural livelihoods are sustained by authorized as well as unauthorized use of about 60 percent of the total geographic area, comprising both cultivated and uncultivated lands. Sixty-seven percent of Uttarakhand's total geographic area, representing most of the uncultivated commons, stands legally notified as "forests" (Ghildyal and Banerjee, 1998). Unlike many other states, all forestland is not under the jurisdiction of the forest department. About 69 percent is reserve forest under the forest department's control, about 16.8 percent is protected forest under the revenue department's jurisdiction (called civil land in the Kumaon region and *soyam* land in the erstwhile state of Tehri Garhwal and managed by the revenue department in collaboration with elected village *panchayats*) and 13.6 percent consists of legally notified village forests managed by elected *van panchayats*. Although socioeconomic differentiation has increased over the years, village communities are still relatively homogenous compared to the high social stratification existing in the plains. Land distribution is relatively equal with rare cases of land holdings of over 2 ha, and landlessness is low.

Women own or control little of the privately held land, and their forest rights are mediated through the male household head in whose name they are recorded. The agro-pastoral economy of the region is still predominantly subsistence based with about 50 percent of rural households, including the rural elite, having high dependence on village commons and forestlands. High migration of men in search of employment makes the women effective managers of the rural household economy. About 40 percent of the households are estimated to be headed by women (CECI, 1998, quoted in Ecotech 1999). The gender-based division of labor is highly skewed, with women responsible for most agricultural work (barring ploughing and marketing), livestock care (excepting during seasonal migration and marketing) and collection of firewood, fodder, leaf litter and non-timber forest produce from village commons and forestlands.

The history of state appropriation of the uncultivated commons has had a profound impact on reshaping traditional gender relations in the area. Prior to the British conquest in 1815, community institutions of the hill peasantry effectively exercised direct control over the use and management of both cultivated lands and the uncultivated commons within customary village boundaries, with little interference from earlier rulers (Agarwal, 1996; Guha, 1989; Nanda, 1999; Somanathan, 1991). Agriculture and animal husbandry comprised inseparable components of hill farming systems, dependent on spatially and temporally integrated use of cultivated and uncultivated lands. Seasonal migration to alpine pastures and grasslands prevented resource degradation by dispensing with the need for uninterrupted use. High dependence on the forests generated conservation values embedded in cultural and religious traditions, such as the maintenance of sacred groves. Traditional village *panchayats* dealt with community affairs and inter and intra-village disputes (Guha, 1989). Despite women being the primary users of forest resources, tradition excluded them from political decision-making at the community level.

A number of interventions during colonial rule permanently altered this landscape of integrated local resource use and management, initially in Kumaon and British Garhwal which were under direct rule, and then subsequently even in the adjoining princely state of Tehri Garhwal. In 1823, the colonial regime undertook the first land revenue settlement in the Kumaon region, categorizing the land within customary village boundaries as cultivated *naap* (measured) and uncultivated *benaap* (unmeasured) lands. In 1893, all unmeasured "waste" (*benaap*) lands in Kumaon were declared "district protected forests" under the control of the district commissioners. From 1910–17, the colonial government attempted to appropriate further control over forest resources by notifying over 7,500 km² of the commons in British territory as reserve forests, thereby severely restricting people's use rights. Peoples' "rights and concessions," such as for grazing and lopping for fuelwood and fodder, so essential for sustaining the agro-pastoral economy, were severely curtailed. The annual practice of burning the forest floor for increasing grass yields was also banned within 1 mile of reserve forests.

As few settlements remained at such distance from the new reserves, it virtually made the practice illegal (Guha, 1989).

Struggles for Local Forest Rights

Forest reservation led to large-scale rebellions and incendiarism (literally setting the reserve forests ablaze in protest against denial of traditional access to them) as it played havoc with the customary patterns of resource use, dislocating existing agrarian practices. The report of the Kumaon Grievances Committee[1] which was set up by an alarmed administration to look into the causes of unrest among the people contains perceptive insights into the prevalent gendered nature of resource use and gender relations. It identified the 'employment of forest guards to enforce numerous rules and regulations and their constant interference with women and children who, under the customs in vogue in Kumaon, are the chief people to exercise on behalf of the villagers such rights as lopping, collection of minor forest produce, grazing, etc.' (GOUP, 1922) as a major grievance. Lopping restrictions imposed in the new reserves accounted for 75 percent of the recorded offences. Although there was no *purdah* in the hills, the men bitterly resented women being summoned to court for such "offences." Men's sentiments against the forest guards' daily interference with women's use of the forest caused the Grievances Committee to recommend that "the forest guard be removed so as to do away with the real grievance in all classes of forests where it can possibly be effected" (ibid.). As a consequence, 4,460 km² of the less commercially valuable forests out of 7,500 km² of the new reserves were taken away from the forest department and handed back to the revenue department with full restoration of people's forest rights. However, rights in these Class I reserves were given to "all bonafide residents of Kumaon," thereby converting customary common property resources into open access areas. Provision was made for *van panchayats* to exercise community control over legally constituted "village forests" demarcated from within the Class I reserves and civil forests, though applicable only in those villages which applied for them. This enabled sections of the peasantry to retrieve some space for local forest management. Thus, by the early 20th century, the uncultivated commons had been divided into three legal categories of forests: commercially valuable Class II reserves under the forest department; commercially less valuable Class I reserves; and civil/*soyam*[2] forests, under the civil administration. This, however, did not appease the villagers' discontent over curtailed resource rights, and protests continued.

The independent state of India continued commercial forest exploitation with even greater vigor. The reach of the forest department and its contractors spread to the remotest corners with the expansion of the road network. Local livelihoods received even less attention than under colonial rule as state policy consistently favored export of raw timber and resin for processing by large industry in the plains. By the 1970s, the Chipko Movement had emerged to demand that priority be given to local employment in the extraction and processing of forest produce (Guha, 1989). Increasing incidents of landslides and floods, and the declining availability of biomass for subsistence needs propelled even the hill women into the movement, broadening the popular base of Chipko protests. Ironically, the Kumaon *van panchayat* rules of 1931 were revised in 1976 at the height of the Chipko Movement, substantially reducing *van panchayat* authority and entitlements even over village forests.

The issue of local forest rights, however, was soon subsumed within the new national and global ideology of environmental conservation. Instead of giving priority to local forest-based livelihoods and employment, Chipko was used to justify a spate of centralized environmental policies and laws. The Forest Conservation Act of 1980 empowered the central government to make decisions related to the alienation of even the smallest patch of forestland. The Uttar Pradesh Resin and Forest Produce Act, 1976, made tapping, sale, and purchase of all resin a state monopoly. In 1981, a 15-year ban was imposed (and has since been extended) on all commercial felling in the Uttar Pradesh Himalayas above 1,000 m.[3] Today the only permitted fellings are for the villagers' timber rights (*haq haquque*) based on consumption recorded in 1911–17 which have not been revised since.

Despite extensive state appropriation and continuing centralization of control over forests, Uttarakhand today provides extensive examples of a unique combination of officially constituted and informal community forest management (CFM) institutions, the latter increasingly developed by women's groups. Autonomous community management of legally demarcated "village forests" (on forest and revenue department land) by *van panchayats* has existed in Uttarakhand for over seven decades. Traditionally practicing restricted democracy due to their almost exclusive control by men, some *van panchayats* have been undergoing internal transformation with women appropriating control over their management. Unofficial community management, with diverse institutional arrangements on all legal categories of forestlands, has co-existed with formally constituted *van panchayats*, and in fact predates them from precolonial times.

As the largest custodian of state property,[4] the forest department has been unable to maintain the forests in good condition or meet people's forest-based livelihood needs. Its responsibility for enforcing the Forest Conservation Act and Wildlife Protection Act has reinforced its image as an antipeople agency. Thus, in 1988–89, some of the Chipko activists who had spearheaded the movement against commercial forest felling, felt compelled to start yet another, relatively less known *ped kato andolan* (cut the trees movement). They argued that the Forest Conservation Act "was being used to hold up basic development schemes for the hill villages while the builders' mafia continued to flout it brazenly under the guise of promoting tourism" (Rawat, 1998: 128). More recently, resource displacement and loss of livelihoods caused by expansion of the protected area network over 20 percent of the geographic area of Uttarakhand, has produced the *jhapto cheeno andolan* (snatch and grab movement) reflecting the intense feelings of alienation and disempowerment. Women who earned international fame for stopping contractors from felling their forests during Chipko have come to hate the word *paryavaran* (environment). As one of these women from Reni village complained: "They have put this entire (surrounding forest) area under the Nanda Devi National Park. I can't even pick herbs to treat a stomach ache any more" (Mitra, 1993: 35).

Both the *van panchayat* and unofficial CFM systems now confront challenges from new systems being imposed on villagers by the state. Under a World Bank funded Forestry Project, the Uttar Pradesh forest department has promoted village forest joint management (VFJM) with autonomous *van panchayats*. This is in contrast to the joint forest management (JFM) practiced on degraded reserve and protected forests under the forest department jurisdiction in other states. It is misleading to refer to the UP VFJM approach as JFM. This hides the crucial difference from JFM in other states—that *van panchayats* had enjoyed autonomous authority over forests prior to VFJM. Reserve forest areas are also to be included in VFJM but till the time of this research (2000), it had primarily focused on *van panchayat* and civil/*soyam* lands (CDS, 2000). VFJM is thus creating space for the forest department to intrude on village forests managed by community institutions instead of creating space for villagers to participate in management of reserve forests under departmental jurisdiction. The decision-making autonomy of *van panchayats* participating in VFJM is now "subject to the supervision, direction, control and concurrence of the Divisional Forest Officer" (FDUP, 1997: 3.1). A functionary of the forest department is being made the joint account holder and the proposed member secretary of *van panchayats* (GOU, 2001), after having no role for seven decades.

At the same time, informal community management is under pressure from the state-directed and target-driven new *van panchayat* formation. The revenue department is demarcating civil forestlands under its jurisdiction as village forests to be managed by officially constituted *van panchayats*. The department is also dividing existing multivillage *van panchayats* without consulting them into single village ones, often generating intervillage inequity and conflict in the process. "Eco-development committees" promoted by the wildlife wing of the forest department solicit villagers' "participation" in replacing their existing forest dependent livelihoods (against their will) with new, nonforest-based alternatives. Decisions earlier taken by villagers through negotiation and consensus-building, such as whether to take up community management at all, whether to do so officially or unofficially, and whether to do so at a hamlet, village or multiple village level, are now being taken by the state on their behalf through the formation of new *van panchayats*. Non-government Organizations (NGOs) and civil society groups, which earlier played an important role in policy advocacy and spearheading social movements, have largely been co-opted to work as "private service providers" for the many donor funded projects in the region.

Women in Informal Community Forest Management

Community forest management outside any formal legal framework is widespread in Uttarakhand in all categories of forestlands, within or near villages. Traditional informal community institutions (*lath panchayats*), informal forest committees (*van samitis*) and more recently, increasing numbers of Mahila Mangal Dals (women's welfare associations) are regenerating and regulating use of reserve and civil/*soyam* forestlands, often compelling unofficial cooperation by forest and revenue department staff. These informal community management systems represent another important form of appropriating space for local forest management.

Women's ability to appropriate such forest management space outside any formal policy framework is shaped by diverse influences and factors. These include the degree of local scarcity of forest resources, and therefore the conflicts and competition governing access to essential forest products; the extent of migration of men from the area; women's exposure to and participation in social movements such as Chipko or empowerment programs such as Mahila Samakhya; support of progressive local male leaders and the indirect impact of positive discrimination policies such as the reservation of one-third seats for women in *panchayati raj* institutions or the promotion of all women *van panchayats*. Out of our 16 case studies, we found village women actively engaged in informal CFM in eight cases, either exclusively through women's groups or together with men.

Holta, one of our case study villages, is a large village of 400 house-holds in Tehri Garhwal district which has a rich tradition of informal CFM. In part, this is because the *van panchayat* rules were not made applicable to the area which was under the erstwhile state of Tehri Garhwal till 1991. Due to not having a *van panchayat*, Holta village initiated protection of its *sanjaiti* (communal) land around 1986 entirely on its own. The area has a high dependence on agriculture with limited migration of men. Literacy rates are low and most men work in the village itself or for wages, or in small shops on the road-side in the vicinity. Although there is visible dominance of men in the village, the local self-government institution (*gram sabha*), is headed by a woman. Interestingly, Holta's informal forest protection commit-tee had no women representatives when it was first constituted 13 years ago. However, despite the *samiti* employing a guard for protec-tion, women of neighboring villages as well as those of Holta itself continued stealing firewood and grass from the forest. According to the men, their failure in stopping the women resulted in them induct-ing four women as members of the committee about five years ago to "get thieves to nab other thieves" (Gairola, 1999a).

Interestingly, the village young men had applied for forming a *van panchayat* six years ago, but had received no response from the administration. Asked why they wanted a *van panchayat* when their informal system was working so well, the men felt that a *van panchayat* had greater access to government funds for plantations. They had heard about generous budgets for VFJM. The women, in contrast, did not want any funds or government scheme. They were proud of their regenerated forest from which they could meet their biomass needs.

On being told that the government was promoting the formation of all women *van panchayats* and that if they submitted such an application it was likely to be cleared quickly, the women's faces lit up. However, the men were taken aback. They said that the village women were illiterate and knew little, and that they would not be able to handle the work. It was then pointed out that the *pradhan* (village head) of Holta *gram sabha* was a woman who was also non-literate. If she could be a *panchayat pradhan*, why couldn't the women manage a *van panchayat*?

According to Holta villagers, water sources had dried up and firewood and fodder had become scarce because of unregulated forest use by the surrounding villages, and encroachment on the common land by local families. Some village youth successfully persuaded the encroachers to vacate the commons, setting an example by giving up their own encroachments. Letters were sent to the *pradhans* of surrounding villages that anyone entering the forest would be fined. Major conflicts followed with one village going to court against Holta due to unclear boundaries of their respective common lands. However, with improvement in forest conditions and availability of water, resistance declined.

At the time of our research in 1999–2000, the entire village's biomass needs, except timber, were met from the regenerated forest. Vegetable cultivation had become feasible with regeneration of three natural water sources. Rules were framed and strictly enforced for grass, tree leaf fodder and firewood collection, with all households contributing to pay the guard. The informal forest protection committee has representation from all the hamlets and castes, and four women representatives of the village Mahila Mangal Dal, empowered by the government's Mahila Samakhya program, had also been able to wedge their way in. Community relations with the forest department, however, were extremely sour. In the words of the village women, "the forest department has made us into thieves. The women were protecting their forest like their own children" (Gairola, 1999a).

Women and men had different perspectives on gender-specific changes in the forest protection committee (*samiti*). According to the women, after gaining exposure and self-confidence through the government's Mahila Samakhya program, they had had to fight hard to gain representation in the protection committee. Due to the limited migration of men from the area the women had few opportunities to occupy leadership roles. The men had resisted their demands saying that they were protecting the forest for the benefit of the women and that there was no need for the women to become members themselves. The men also considered it difficult to decide which caste groups the women should represent as the men of all castes considered themselves indispensable. The women insisted that if seats were now being reserved for them even in *panchayats*, the same could be done for them in the village forest protection committee (*samiti*). Once two of the women were elected as ward members during the *panchayat* elections, they were made members of the forest protection *samiti* (FPS) also. Significantly, most members of the FPS are also ward members of the village *panchayat*.

According to the women, now all their grass, firewood and bamboo requirements are met from the forest. Similarly, all the villagers' small timber needs for agricultural implements, such as sickles, ploughs and axes are met from the forest. Everyone is satisfied with the work of the forest protection *samiti*. Daily life has become much easier for their daughters and daughters-in-law. Although a few troublesome individuals do keep creating problems now and then, it is easy to keep them quiet as the women of their families too are benefiting from easy availability of firewood and fodder. When asked whether the villagers had helped put out fires in government forests during the summer, the women's response was:

Why should we help when those are not our forests? The Forest Department does not permit us to lift even a dry branch from them. After the government forests are burnt, we are able to fetch firewood from them. So we actually benefit from fires in government forests. We have not taken a contract for environmental protection! If there is environmental damage, it should equally affect the Forest Department. They are also citizens of this country but behave as if they are the descendants of the British! (Gairola, 1999a)

Such antipathy towards the forest department staff among village women is widespread. Once, the women of Dhar Kot village in Pratapnagar block of Tehri district tied the forester and members of the forest patrol to trees and continued lopping green oak leaf fodder for their cattle. The forest staff kept shouting the whole day in the thick forest, but no one was there to hear them. While most village men are fully aware of their rights and the procedures for exercising them, the women knew next to nothing about such matters.

In several hamlets of the three villages under Makku the *van panchayat*, Mahila Mangal Dals had asserted informal control over patches of civil or communal land closer to their settlements for day to day management of firewood and fodder. According to the former *sarpanch* of the Makku *van panchayat*, the scarcity of firewood and fodder is increasing conflicts over forests to such an extent that women have started resorting to physical fights among themselves. He had encouraged the village women to enclose patches of communal *gram panchayat* land for meeting their needs, while also saving it from encroachment by the elite. Both he and the women faced a lot of resistance from powerful vested interests and husbands who had to do housework while women patrolled. However, effective protection by the women led to dramatic regeneration of the women's forests/ *mahila bans* (Bhatt, 1999).

In Bareth village, which also has a *van panchayat*, while the men control the pine dominated *van panchayat* forest, the women effectively manage and control the mixed species forest on the village common land. The forest is closer to the village and useful for meeting their daily requirements of fuelwood and fodder. In both Makku and Bareth, women perceived local *van panchayat* councils to be dominated by men. *Panchayat* forests were also far from the villages, and therefore not convenient for daily fuelwood and fodder collection. The formal and informal CFM arrangements here complemented each other with the women occupying informally carved out space. They could access such space with the mediation of the *gram panchayat* without having to deal with cumbersome bureaucratic procedures. In Arakot village, the Mahila Mangal Dal has been protecting the village *soyam* land for the past 20 years, paying a guard with voluntary contributions. In Naurakh and Resal, civil land was being protected by individual families through private enclosures with day to day management under the women's control. Officially viewed as "encroachment" on government lands, such informal systems are fairly widespread owing to their low transaction costs (Singh, 1997a).

Women Appropriating Space Within Van Panchayats

Till the 1970s, *van panchayat* councils seem to have been almost exclusively male due to the traditional exclusion of women from community institutions. The Chipko Movement for the first time drew women out from their homes exposing them to a new world of assertion and articulating their demands. The movement also had a significant impact on traditional gender relations, with the men accepting women's participation in nondomestic affairs, particularly those related to management of forest resources. The subsequent period has also seen specific government policies promoting women's participation in forest management, including promotion of women's representation on *van panchayat* councils. Our case studies included one case where government officials had compelled a woman to be elected as a *van panchayat sarpanch* (in Bareth village) simply on the grounds of it being government policy. In the absence of any capacity building support, or any previous exposure to such a role, her forced election to the position of sarpanch was a disaster. She resigned within a month.

In two other cases, however, women had appropriated space for themselves within their respective *van panchayats* to have a decisive say in their day to day functioning. In the case of the Pakhi *van panchayat*, the Mahila Mangal Dal had gained informal control over the *van panchayat* whereas in the case of Dungri Chopra *van panchayat*, the women had succeeded in getting an all women *van panchayat* council elected. Pakhi *van panchayat* had subsequently been brought under VFJM under the ongoing World Bank funded forestry project. We examine here how women appropriated space for themselves in the two *van panchayats* and their functioning under the women's leadership.

The Pakhi *van panchayat* was formed in 1958 and it lies in Chamoli district, in the vicinity of Gopeshwar from where Chandi Prasad Bhatt spearheaded the Chipko Movement in the early 1970s. The village became well known for the active participation of women and men in the movement which stopped the felling of their forests by the forest department. Pakhi and Jalgwad, the two villages falling under the same *gram panchayat,* have a common *van panchayat*. Out of about 180 total households only eight are scheduled castes and the rest are upper caste Hindus (Thakur or Brahmin). The *van panchayat* forest area of 240 ha is in good condition and has mixed species dominated by oak and rodhedendrum, with a sprinkling of deodar (blue pine). Fuelwood, fodder, animal bedding and some non-timber forest produce, rather than cash income, are the primary benefits the villagers get from the forest. There is limited male migration from the area.

The two villages have an active Mahila Mangal Dal, whose leaders have received considerable exposure since their involvement with Chipko. The Mahila Mangal Dal continues to interact with several

NGOs, participate in government-sponsored camps for women and is also a member of Himvanti, a multicountry federation of mountain women's organizations for the Hindukush Himalayan region. In 1999–2000, although the elected council had two women members, they did not participate actively in *van panchayat* meetings. Instead, the Mahila Mangal Dal as a whole influenced the *van panchayat* council by the women's collective decisions. One of the elected women *van panchayat* members had received training and marketing support from Himalayan Environmental Studies and Conservation Organisation (HESCO), an NGO promoting economic development in villages, and has initiated processing of locally grown fruit into jams, pickles, and juices. This has increased the villagers' returns from fruit significantly.

Participation in the Chipko Movement enabled the Mahila Mangal Dal to effectively wrest control over the day to day management of the village forest from the male dominated *van panchayat* council. Prior to the initiation of VFJM with the *van panchayat* in August 1999, decisions about when to open the forest for grass, leaf and firewood collection, the rules for collection, the fines for violation, etc., were taken by the Mahila Mangal Dal and communicated to the *van panchayat sarpanch*. His primary responsibility was to publicly announce these decisions in both the villages. As no external funds were available, the women also used to repair the forest boundary wall with voluntary labor. They had employed a woman as forest guard to whom they paid Rs 300 per month also raised through voluntary contributions. Fines from those violating the protection rules were collected by the Mahila Mangal Dal and deposited in its own account. At the time of the research, the Mahila Mangal Dal had Rs 3,100 in its account.

The women's control over forest-use decisions enables them to ensure that forest product collection does not conflict with periods of heavy agricultural work. Soon after harvesting the monsoon *mandwa* crop in October, they open the forest closest to the village for grass collection. The furthest forest area is opened in December when all agricultural work is finished and the women can devote most of their time to stocking up firewood and grass before the snow falls. Cutting of bushes and pruning tree branches is done from April to May.

Although pleased with having appropriated control over the village forest, the women expressed resentment over the men leaving all the forest protection work to them. They had attempted to coax the men to assist with voluntary patrolling but the men had refused saying that it was only the women who needed the forest. The women also reported that when outsiders came to the village (and there are many visitors due to the contacts established during Chipko) the men push the women forward to talk to them. However, when important village related decisions are made, the women are often kept in the dark. This was evident with the introduction of the VFJM with the *van panchayat* in August 1999. Despite the World Bank project document's emphasis on specific targeting of women and the poor in "participatory" forest management, neither the forest department-NGO "spearhead team" nor the village men provided the women much information about VFJM. The men only told them that a budget of Rs 2 million was being approved for the *van panchayat* which would be very beneficial for the village.

The sudden availability of a generous budget for the village forest, however, led to a rapid gender-based shift in power and control. The same men, whom the women had complained that they had left all forest protection work to them, suddenly became overenthusiastic for it. Three watchmen, at salaries of Rs 1,000 per month, were employed together with one watchwoman for forest protection. After three months of working without a salary, the *sarpanch* started paying the woman Rs 200 per month. Knowing that the men were being paid Rs 1,000 per month, she refused to accept such a payment. After a lot of arguing, she was finally paid Rs 700 for the previous month's work and then laid off on the grounds that it was difficult for a woman to protect the far ends of a large forest. Similarly, initially the men monopolized wage work in the nursery. Only after strong protests by the women were some women also employed. When no funds were available for forest development, women were left to take care of it with voluntary labor. As soon as money came in, the women were labeled ineffective for undertaking the task.

The men too, however, were not outright winners in the subtle shifts in the balance of power and control within the village. The *van panchayat* council and the *sarpanch* experienced a similar loss in local decision-making control to the forest department. Maintenance of the muster roll for wage work and preparation of the monthly progress report was now done by the guard or forester instead of the *sarpanch* as earlier. According to the *sarpanch*, the villagers' role in VFJM was reduced to providing information for preparation of the microplan and working as paid labor for forestry operations. The villagers could no longer do anything on their own without prior approval of the forest staff. Neither the women nor the men were clear about the new VFJM rules, or the legal agreement they were supposed to have signed. There was no copy of the agreement in the *van panchayat* records, and the *sarpanch* did not have a copy of the microplan with him. He said that years of experience had made him familiar with the rules governing *van panchayats*. But he knew little about the VFJM rules.

In the words of one of the worried women, in their greed for money, the men had made a deal over their village forest with the forest department. Since conducting the fieldwork an all women *van panchayat* council has been elected in Pakhi. However, the husband of the woman *sarpanch* is the forest guard who is the joint signatory for the *van panchayat's* VFJM account with his wife as the *sarpanch*. Due to this, her questioning voice related to VFJM has become muted. It is not known how this has impacted women's unity within the Mahila Mangal Dal.

The Dungri Chopra *van panchayat* in Pauri Garhwal district was formed in 1939 and is the oldest *van panchayat* among our research case studies. The district has a high literacy rate combined with high rates of migration of men. Due to the men getting good jobs, their interest in managing the village forest has declined which has left the village women with problems resulting from poor satisfaction of their biomass needs. Under the Mahila Samakhya program the women have gained self-confidence, are better organized, and better informed about government policies for women's empowerment. They made a concerted bid to take over the management of the village forest from the disinterested men's leadership. Many men disapproved of the women's initiative, but some of the village elders encouraged the women to take over. As a consequence, around 1997–98, the village women succeeded in getting an all-woman *van panchayat* council elected. The woman *sarpanch*, however, faces daunting challenges.

Today, government schemes worth millions come to the villages and there is rampant corruption. No government official visits the village without negotiating a commission in advance. In 1999, the District Rural Development Agency sanctioned Rs 60,000 for undertaking plantation in the village forest. When Dwarka Devi, the woman *sarpanch*, went to collect the first installment of Rs 30,000, the *van panchayat* inspector made her sign a receipt for the full amount but gave her only Rs 24,000. She went to Dilip Singh, an elder former *sarpanch*, to seek advice on what to do. He told her that in future, whenever any such payment had to be collected, she should always take other women *panches* with her, and on returning to the village, place the entire amount in front of the general house to prevent anyone from suspecting her. The villagers would themselves help her work out how to deal with the situation (Gairola, 1999c).

Seemingly, Dwarka Devi had internalized this valuable lesson in transparent governance. It has enabled her to maintain collective responsibility for managing the village forest and also to evolve coping strategies for dealing with the increasingly unsavory and dramatically changing world outside the village. The *panchayat* forest is one of the best in the district and the women meet almost all their forest needs from it.

Rather than strengthening such transparent governance mechanisms within *van panchayats*, VFJM, which is being promoted under the World Bank funded forestry project, assumes that misuse of funds for implementing microplans as well as the misuse of the villagers own funds can be prevented by making a forest department functionary as the member secretary-cum-joint account holder of the *van panchayats*. While perverting the tradition of the leadership's accountability to the general body of villagers, the arrangement has created yet another avenue for lower level forest department staff forging alliances with a class of elite village men to misappropriate microplan funds.

Women-Specific Impacts of VFJM: Livelihoods and Equality

Discussions on the merits and demerits of VFJM rules often center around the percentage of the share of income that the villagers would get from their forests. Women forest users, however, have been driven to physically attack other women competing for increasingly scarce fuel and fodder resources, some even resorting to suicides to end their daily drudgery (Nanda, 1999). Their priorities are to increase the direct use values of their forests. An improved quality of life and ecological security for them precedes considerations of income from sale of forest products, although additional income is never unwelcome. The project document claims to target women and the poor but provides no analysis of how a shift in management priorities for increasing income would impact their access to requirements for daily subsistence or their work burdens. VFJM microplans in the case studies had few provisions for supporting livestock, a critical local livelihood requirement, instead emphasizing closure of forests to grazing.

A key assumption underlying support for VFJM is that the major problem plaguing *van panchayats* was the lack of funds, and therefore the best incentive for increasing the villagers' stake in forest protection was offering them attractive shares of income from the sale of forest products. Yet, a survey of 644 *van panchayats* in Ranikhet sub-division in Almora district found that as many as 433 did not have any income and only 45 could boast of a balance of at least Rs 25,000 in their accounts (Singh and Ballabh, 1991, quoted in Singh, 1997b). A large number of forests had been managed well by villagers without any source of income. It has also been observed that the income of a *van panchayat* has no bearing on what the villagers consider to be a "good" *van panchayat*. A *van panchayat* is regarded as good if it meets the needs of fuel and fodder and helps recharge water sources. Therefore, oak forests were generally preferred to pine forests even though they provided less revenue and employment (through resin tapping). In contrast, the revenue department considered *van panchayats* with bigger bank balances to be performing better. The World Bank forestry project subscribed to the same assumption.

The World Bank funded forestry project has provided an average of Rs 1.5–2 million for implementing a microplan in each village brought under VFJM under the project. Besides promoting inequity between neighboring villages, the sudden offer of large sums of money to selected villages with high unemployment and limited opportunities for cash incomes, however, had led to the eruption of major conflicts to gain control over the funds. Sustainable voluntary protection, often by womens' groups, had been replaced by patrols of externally-funded guards. In our case study villages, selection of the paid guards had been done by elite village men providing them a new avenue for patronage. Women's groups, in particular, had been able to negotiate/assert their authority to manage civil/*soyam* lands by negotiating with their *gram panchayats*. In at least three out of 10 case studies (Pakhi, Arakot, and Chora) where VFJM was introduced, village women were actively protecting the *van panchayat/soyam* forests. In all three cases, no effort was made to build upon, and strengthen the women's efforts. In Pakhi village, a poor watchwoman paid by women's voluntary contributions was replaced by four watchmen paid much higher salaries with World Bank loan funds.

While overlooking existing systems of voluntary contributions, the project demands that villagers contribute 20 percent of microplan costs. This was being collected through compulsory deductions from wages, thereby transferring the costs to those doing the wage work. The majority of these workers were women or poorer villagers. They had been forced to contribute on behalf of the whole community, often without even being aware that there were deductions and as to why they were being paid less than the minimum wage. In none of the case study villages had any open discussion been held on how the mandatory contribution could be shared more equally by all those theoretically gaining entitlements to the specified benefits. In Kharag Karki village, women thought they could at least take the firewood from cleaning operations as compensation for accepting lower wages, but were not permitted to do so, leaving them alienated and bitter. Organized and acutely forest-dependent women have borne disproportionate costs of (in)voluntary contributions or unpaid protection duties in order to build up *panchayat* and forest committee funds controlled by elite men and the forest department staff. This situation is highlighted by recent developments in the much publicized case of the Parvera *van panchayat* which has had an active woman *sarpanch*. The department had encouraged the women to take up voluntary protection to build up the *van panchayat* fund with the project money available for protection. This is considered an indicator of good *panchayat* functioning under VFJM (FDUP, n.d.). However, with the transfer of a committed senior forest officer who had been taking personal interest in supporting the woman *sarpanch* and the Mahila Mangal Dal, the good performance of the Parvera *van panchayat* has proved to be short lived. Elite village men have made a bid to appropriate control over the women's savings in the VFJM account by getting a new *van panchayat* elected in cahoots with the *van panchayat* inspector of the revenue department. The woman *sarpanch* has challenged the new election in court on the grounds that a new council cannot be elected during the five-year agreement for VFJM signed by the existing *van panchayat* with the forest department. While the old and the new *sarpanches* are fighting the battle in court, the biggest casualty has been the well protected village forest due to the disputed authority over its management (Author's interview, June 22, 2001).

Conclusion

In Uttarakhand, NGOs and civil society groups have historically played a strong advocacy role. Chipko, for example, was triggered by protests led by Dasholi Gram Swaraj Mandal. Today, the NGO movement is split into different camps and factions. The vast majority have been co-opted to work as "private service providers" for the several donor funded projects in the region, including the forestry project. Once they have accepted working on project terms, they effectively lose their critical and questioning voice. Among civil society groups perturbed over such impacts of donor funded projects, there are different sub-sets and worldviews. The overall impact is that today the NGO and civil society movements have been considerably weakened with hardly any concerted public action for protecting people's forest rights. Consequently, no unified voice has been raised against the potential damage to the region's unique institution of the *van panchayat* from the VFJM. A large number of concerned individuals and advocacy groups, however, have been articulating such concerns at different forums (SKS, 1999; SPWD, 2000).

A series of government actions has weakened the existing local forest management systems substantially over the course of the last century. Progressive restrictions on local use of forest resources through the Forest Conservation Act, the felling ban, and recent Supreme Court judgements, combined with large-scale livelihood and resource displacement caused by expansion of the Protected Area network, are changing people's attitudes towards forests and undermining the primary incentives for CFM.

However, despite the imposition of crippling bureaucratic controls on their functioning, a large number of Uttarakhand's *van panchayats* have survived as vibrant self-governing community forestry institutions. A large number of diverse and informal institutional arrangements for community management on all legal categories of forestlands, several of them led by acutely forest-dependent women, co-exist with, and even within, the formal *van panchayats*. Such informal arrangements, often with negotiated support of elected *gram panchayats*, many now headed by women, provide more accessible and flexible space for CFM by poor women and marginalized groups outside the ambit of bureaucratic procedures and controls. These formal and informal community institutions have traditionally functioned on principles of restricted participatory democracy by men. The larger context of sociocultural transformation and change, however, is creating space for women and other marginalized groups to broaden their democratic base. Women have started asserting their rights to participate in community decision-making and defining forest use and management priorities through organized action and struggle within their households and communities.

In the name of devolution, VFJM is empowering the forest department to reassert control over both *van panchayat* forests and civil/*soyam* lands, the only surviving village commons. Instead of revalidating the rich and diverse base of indigenous knowledge of local women and men, and the diverse management systems they have developed for supporting livelihoods and maintaining ecological services, VFJM reinforces the forest department's claim to be the monopoly holder of technical forest knowledge, despite its historical lack of experience with forest livelihoods and biodiversity conservation. Externally imposed microplanning teams are insensitive to the internal dynamics of existing self-governing institutions and women's ongoing struggles within them for gaining greater voice and control over livelihoods and decision-making. Instead of empowering women, such top-down interventions do the opposite by disrupting and marginalizing women's own struggles and initiatives. Placing a forest department functionary as the joint account holder and proposed member secretary inside *van panchayats* shifts institutional accountability to the forest department and away from forest users, of whom a great majority are women.

Nurturing democratic, self-governing CFM institutions requires a framework ensuring tenurial security over community forests, clear boundaries defining communal property rights and empowerment of forest-dependent women and men within communities to make real choices for enhancing sustainable livelihoods in accordance with their gender-specific priorities. State interventions need to build upon and further facilitate gender equal democratization of existing local initiatives and institutional arrangements. These need to take into account women's traditional lack of independent access and control over communal forest resources and decision-making in forest management institutions, instead of seeking to replace them wholesale with standardized state-engineered institutional frameworks such as VFJM.

Further support for policy interventions such as VFJM, target-driven new *van panchayat* and Eco-development Committee formation in Uttarakhand needs to be abandoned. Alternatives to these state-sponsored institutional structures need to be developed through broad based consultative processes in which the least powerful and most forest-dependent, women in particular, are assured prominent opportunities for discussion and decision-making.

NOTES

This article is based on research funded by the Centre for International Forestry Research (CIFOR) and International Fund for Agricultural Development (IFAD). I am grateful to Geeta Gairola, Tarun Joshi, other *Sainion Ka Sangathan* team members and Seema Bhatt for their field work and to Govind Kelkar for comments and editorial inputs. The article has drawn considerably on two of my earlier papers—a draft overview paper based on the research in Uttarakhand titled 'From Right Holders to 'Beneficiaries'? Community Forest Management, Van Panchayats and Village Forest Joint Management in Uttarakhand', January, 2001 and 'Disempowerment in the Name of 'Participatory' Forestry?—Village Forests Joint Management in Uttarakhand', Forests, Trees and People Newsletter No. 44, April 2001, Uppsala, Sweden.

1. The Forest Grievances Committee for Kumaon set up in 1921 examined some 5,040 witnesses from all grades of society in British Garhwal, Almora, and Nainital.
2. In the adjoining Tehri state, nonreserve forestlands under the civil administration were called *soyam* lands.
3. In 1986, the ban was made applicable above an altitude of 2,500 m. At lower altitudes, green felling of only pine in areas specified in forest working plans is permitted (Saxena, 1995).
4. Out of 67 percent of Uttarakhand's area classified as forests, about 69 percent is reserve forests exclusively under the jurisdiction of the forest department (FD). The rest, comprising of civil/*soyam* and VP forests, respectively falls under the revenue department and *van panchayat* jurisdiction with the forest department responsible for technical supervision.

REFERENCES

Agarwal, Chetan (1996) 'Boundary and Property Rights in Uttarakhand Forests,' *Wastelands News*, February-April, New Delhi.

Bhatt, Seema (1999) 'A Case Study of Some People's Institutions in the Akash Kamini Valley,' Garhwal, Mimeo.

Center for Development Studies (CDS) (2000) *Sanyukt Van Prabandhan*, 3(1), Nainital.

Canadian Centre for International Studies and Cooperation (CECI) (1998) 'Community Based Economic Development Project, India: Preliminary Proposal,' CECI, Montreal, Mimeo.

Ecotech (1999) 'Study on Management of Community Funds and Local Institutions, State Perspectives and Case Studies,' Volume II, Draft for Discussion, Ecotech Services, New Delhi, Mimeo.

Forest Department of Uttar Pradesh (FDUP) (1997) Village Forest Joint Management Rules, August 30, Lucknow.

Forest Department of Uttar Pradesh (FDUP) (n.d.) 'Empowerment of People through Forestry—A Status Paper on JFM in U.P.,' Lucknow.

Gairola, Geeta (1999a) Field case study of CFM in Holta Village, district Tehri Garhwal.

———— (1999b) Field case study of VFJM with Pakhi Van Panchayat, district Chamoli.

———— (1999c) Field case study of Dungri Chopra Van Panchayat, district Pauri Garhwal.

Ghildyal, M.C. and A. Banerjee (1998) 'Status of Participatory Management in Uttarakhand Himalayas,' Paper presented at regional workshop on participatory forest management implications for policy and human resource development, Kunming, People's Republic of China.

Government of the United Provinces (GOUP) (1922) Kumaon Forest Grievances Committee Report, Government of the United Provinces, Lucknow.

Government of Uttaranchal (GOU) (2001) The Uttaranchal Panchayati Forest Rules, 2001, Final draft prepared by the UPFD in May 2001, Dehradun.

Guha, Ramachandra (1989) The Unquiet Woods: Ecological Change and Peasant Resistance in the Himalaya, Oxford University Press, Delhi.

Mitra, A. (1993) 'Chipko, An Unfinished Mission,' Down to Earth, April 30, New Delhi.

Nanda, N. (1999) Forests for Whom? Destruction and Restoration in the UP Himalayas, Har Anand Publications, New Delhi.

Rawat, A.S. (1998) 'Biodiversity Conservation in U.P. Hills: A People's Viewpoint,' Studies, 3(4), Centre for Sustainable Development, UP Academy of Administration, Nainital.

Sainion Ka Sangathan (SKS) (1999) Proceedings of NGO discussion held in July, 1999.

Saxena, N.C. (1995) Towards Sustainable Forestry in the UP Hills, Centre for Sustainable Development, LBS National Academy of Administration, Mussoorie, UP.

Singh, K. and V. Ballabh (1991) 'People's Participation in Forest Management: Experience of Van Panchayats in UP Hills,' Wasteland News, Aug–Oct, pp. 5–13.

Singh, Satyajit (1997a) 'Diverse Property Rights and Diverse Institutions: Forest Management by Village Forest Councils in the U.P. Hills,' IDS, Sussex.

———— (1997b) 'Collective Action for Forest Management in the UP Hills,' IDS, Sussex, Mimeo.

Somanathan, E., (1991) 'Deforestation, Property Rights and Incentives in Central Himalaya,' Economic and Political Weekly, 26(4): PE37–PE46.

SPWD (2000) Proceedings of the Seminar on Sustainable Forest Management in Uttarakhand held on February 22–23 at Dehradun.

ABOUT THE EDITORS AND CONTRIBUTORS

THE EDITORS

Govind Kelkar is Coordinator, IFAD-UNIFEM Gender Mainstreaming Programme in Asia, New Delhi, and the founding Editor of the journal *Gender, Technology and Development*. She has previously taught at Delhi University, the Indian Institute of Technology, Mumbai, and the Asian Institute of Technology (AIT) where she also founded the graduate program in Gender Development Studies. Dr Kelkar has previously co-authored *Gender and Tribe* (1991), and co-edited *Feminist Challenges in the Information Age* (2002). She has also contributed numerous articles to scholarly journals with a focus on gender relations in Asia. Dr Kelkar is a frequent consultant to IFAD, Rome, UNIFEM, New Delhi, and other UN organizations on mainstreaming gender in development, and is a keynote speaker at women's conferences.

Dev Nathan is an economist and Honorary Professor, Institute of Ethnology, Yunnan Academy of Social Sciences, Kunming, China. He is also a columnist and regular contributor to the *Economic and Political Weekly*. Dr Nathan has previously co-authored *Gender and Tribe* (1991), *Assessment of Rural Poverty in Asia and the Pacific* (2002), and edited *From Tribe to Caste*. Dr Nathan frequently serves as a consultant to IFAD, Rome, several UN organizations, and ICIMOD, Kathmandu, on issues of rural poverty and the development of indigenous people.

Pierre Walter is Assistant Professor, Department of Educational Studies, University of British Columbia, Vancouver, Canada. Dr Walter's research interests include literacy, immigrant and extension education, comparative education and policy studies, alternative education, Asian studies, and gender and development. He has contributed numerous articles to leading scholarly journals, besides having served as Assistant Editor of *Gender, Technology and Development.*

The Contributors

Yang Fuquan is Professor at the Institute of Ethnology, Yunnan of Academy of Social Sciences, Kunming, China.

Samar Bosu Mullick is a Ph.D. scholar and activist of the Jharkhand movement, Ranchi, India. He is commonly known as Sanjay.

Indra Munshi is Reader, Department of Sociology, Mumbai University, Mumbai, India.

Tiplut Nongbri is Associate Professor, Anthropology, School of Social Sciences, Jawaharlal Nehru University, New Delhi, India

Paul Porodong is Assistant Professor, Sociology and Social Anthropology Program, University Malaysia, Sabah, Malaysia.

Cholthira Satyawadhana is Associate Professor and Dean, Thai Studies Program, Rangsit University, Pathumthani, Thailand.

Madhu Sarin is a scholar and activist in community forest movement, Chandigarh, India.

K.S. Singh is Former Director/General, Anthropological Survey of India. He is the author of *Peoples of India* and numerous other books on the indigenous peoples of India.

Xi Yuhua is Associate Professor, Dongba Research Institute, Lijiang, Yunnan, China.

He Zhonghua is Director of the Indigenous Women's Resource Center Lijiang, Yunnan, and former Professor, Institute of Ethnology, Kunming.

INDEX

mobilization of, 40
opportunities for, 158
patriarchal control over, 260
patrilineal Confucian values of a, 19
persecution of, as witches, 22
pervasiveness of the demonization
    of, 22
political status of, 244
predominance of, 198
prohibitions for, 55
reproductive services of the, 203
role of, 13
role of, as gatherers, 49
shamans, 18, 19
social capital for forest, 37
subordination of, 266
subordination of, in forest
    societies, 25
use of sophisticated arms by, 52

village head in a matrilineal
    community, 165
workloads of forest-based, 13
women's economic, political and
    ritual roles, 14
women's inclusion in committees, 15
World Bank-funded forestry
    project, 304

Xi Yuhua, 20, 22

Yamba, C. Bawa, 275
Yanagisako, S.J., 236
Yang Fuquan, 16, 17, 25
Yang Shengming, 151
Yu Xiaogang, 22, 42

Zahir, Naved, 277, 284
Zilman, Adrienne L., 49